Studies in Computational Intelligence

Volume 1282

The series "Studies in Computational Intelligence" (SCI) publishes new developments and advances in the various areas of computational intelligence—quickly and with a high quality. The intent is to cover the theory, applications, and design methods of computational intelligence, as embedded in the fields of engineering, computer science, physics and life sciences, as well as the methodologies behind them. The series contains monographs, lecture notes and edited volumes in computational intelligence spanning the areas of neural networks, connectionist systems, genetic algorithms, evolutionary computation, artificial intelligence, cellular automata, self-organizing systems, soft computing, fuzzy systems, and hybrid intelligent systems. Of particular value to both the contributors and the readership are the short publication timeframe and the world-wide distribution, which enable both wide and rapid dissemination of research output.

Indexed by SCOPUS, DBLP, WTI AG (Switzerland), zbMATH, SCImago.

All books published in the series are submitted for consideration in Web of Science.

Tarun Kumar Sharma · Sumit Singh Dhanda ·
Deepak Panwar · Brahmjit Singh · Poonam Jindal

Understanding Internet of Things: Evolution, Enabling Technologies and Challenges

Springer

Tarun Kumar Sharma
Department of Computer Science
and Engineering
Shobhit University Gangoh
Saharanpur, Uttar Pradesh, India

Sumit Singh Dhanda
School of Computer Science
and Engineering
IILM University
Greater Noida, Uttar Pradesh, India

Deepak Panwar
Department of AI-ML
School of Computer Science
and Engineering
Manipal University Jaipur
Jaipur, Rajasthan, India

Brahmjit Singh
Department of Electronics
and Communication Engineering
NIT Kurukshetra
Kurukshetra, Haryana, India

Poonam Jindal
Department of Electronics
and Communication Engineering
NIT Kurukshetra
Kurukshetra, Haryana, India

ISSN 1860-949X ISSN 1860-9503 (electronic)
Studies in Computational Intelligence
ISBN 978-3-032-21587-1 ISBN 978-3-032-21588-8 (eBook)
https://doi.org/10.1007/978-3-032-21588-8

This Springer imprint is published by the registered company Springer Nature Switzerland AG
The registered company address is: Gewerbestrasse 11, 6330 Cham, Switzerland

Competing Interests The authors have no competing interests to declare that are relevant to the content of this manuscript.

Contents

List of Figures

List of Tables

Chapter 1
Introduction to Internet of Things

Internet of things (IoT) is the realization of ubiquitous computing paradigm. IoT was envisioned with the idea to minimize the wastage, loss and cost in supply chain management using Radio Frequency Identification (RFID), in 1999 by Kevin Aston [1]. RFID has been used for connecting things. It enabled these things to be identified and tracked and this information was shared over the internet which in turn helped in reducing wastage, loss, and cost. At that time, technology had its limitations and research efforts were needed. Neither small sensors had any technology to connect with internet nor there any existing infrastructure to a huge number of communicating devices. But with the advancement in integrated circuits (IC)-technology and development of Internet Protocol version 6 (IPv6), which can uniquely identify nearly everything on this planet, that vision can be realized today. The IoT objects can sense and do various tasks via information sharing and coordinated decision making amongst each other. The IoT has converted the traditional objects into the smart ones by utilizing the existing technologies [2].

Cisco considers IoT to be the first evolution of the Internet [3]. Identification, sensing, communication, computation, semantics, and services are the six elements of an IoT system. The most important technical advancement that has made it possible is IPv6. It has given the capability of uniquely identifying each and every connected object on earth. IoT has gone through three generations of development till now.

Evolution of IoT has helped in the growth of IoT devices. As per Shanzi et al., [2] in 2014 there were at least 9 billion interconnected devices in China alone, number will touch the 24 billion in 2020. As per Cisco IBSG, IoT is that point in time when 'connected objects' will outnumber the people on earth. In 2003 the people living on the earth were 6.3 billion while the 'connected devices' were 500 million. Cisco IBSG predicted in its report [3] in 2011 that in year 2015 the number of devices connected to Internet will be 25 billion while the same will rise to 50 billion in 2020. The Internet of things paradigm benefits all daily life aspects by connecting any device at any place that too any time [4].

T. K. Sharma et al., *Understanding Internet of Things: Evolution, Enabling Technologies and Challenges*, Studies in Computational Intelligence 1282,
https://doi.org/10.1007/978-3-032-21588-8_1

Wireless Sensor Networks (WSN), cloud computing, Big Data techniques, Software Defined Networking (SDN), Information Centred Networking (ICN) and Fog Computing are the technologies that help in formation of IoT. In this survey we have tried to identify the area of research that can be used by the researchers to work for IoT.

1.1 IoT Architecture and Conceptual Framework

Conceptually, IoT is a system designed to achieve the optimal utilization of resources. Its objective is to minimize loss, reduce wastage, and cutdown cost. This system collects the data/information with the help of different sensors; the data is communicated over the network to cloud where information is extracted from it that information is communicated for the intelligent decision making to different users or applications. In IoT where things have virtual identities, through which these can be identify throughout the world. Virtual identities are used to manage these things throughout the world.

IBM provides the IoT conceptual framework. It is more like the blueprint of the building providing the exact idea what will be the constituents. But at implementation level details can be changed. Everything will be different. How everything will be connected.

The IoT architecture serves as the structure for developing an IoT application. This is a subject on which consensus cannot be achieved. Numerous groups are making attempts, although no singular architecture has gained universal acceptance. The rationale is attributed to the diverse applications of IoT-based technologies. Each application possesses an own set of needs.

Figure 1.1 illustrates six distinct architectures referenced in the literature, including IoT-A, RAMI 4.0, IIRA, Web of Things (WoT), IoT-RA, and Arrowhead. Various organizations offer these architectures to establish a foundation for developing IoT applications. IoT-A is an initiative stemming from the FP7 research project of the European Union. It comprises five distinct models pertaining to domain, information, communication, function, and security. The Industrial Internet Reference Architecture (IIRA) and the Reference Architecture Model Industrie 4.0 (RAMI 4.0) are two frameworks centred on Industry 4.0. IIRA integrates four distinct perspectives and provides five designs derived from these perspectives. Conversely, RAMI 4.0 is a three-dimensional framework featuring a vertical axis and two horizontal axes. The vertical axis illustrates six layers representing the machine's composition. The left horizontal axis emphasizes the life cycle of items and facilities. The right horizontal axis displays the seven tiers of functional hierarchy. The Arrowhead framework, developed by the eponymous organization, comprises a system that integrates eight distinct subsystems. It is a hierarchical structure ranging from level 0 to level 4 [5, 6].

WoT comprises a four-layer design. An examination of various designs reveals that architecture is determined by the demands of use cases. A singular solution

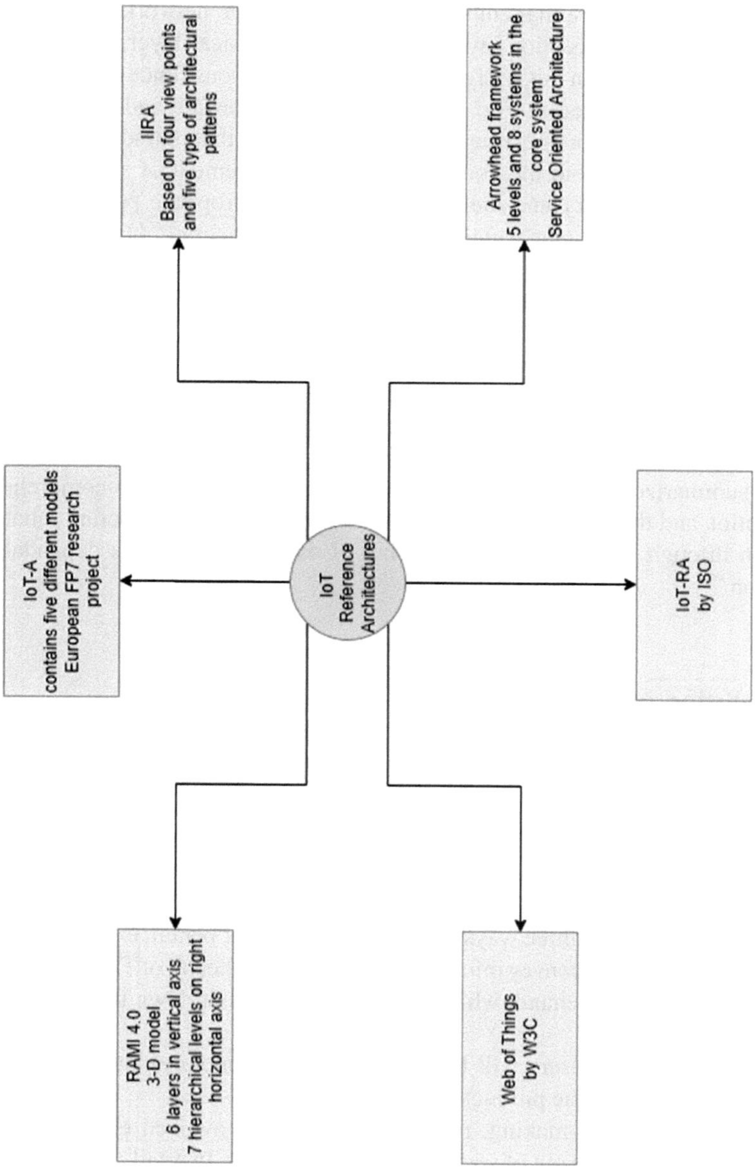

Fig. 1.1 Reference architectures for IoT [5–7]

for all problems cannot be established. Consequently, only essential elements can be articulated explicitly, but a universal architecture applicable to all applications cannot be delineated.

Figure 1.2 illustrates four distinct architectures. Figure 1.2a illustrates a generalized three-tier architecture that emphasizes the application, network, and perception layers alone. The perception layer resembles the physical layer, however the application layer outlined in this design encompasses numerous functionalities. The functionalities are delineated in Fig. 1.2b within the middleware and coordination layer. Figure 1.2c illustrates a service-oriented architecture that delineates the tasks of the application layer from the coordination and management of services. This comprises service discovery and the allocation of the appropriate protocol based on the request. The service management layer encompasses certain functions of the network layer as well. The techniques employed for service discovery are addressed in a subsequent section of this study. Figure 1.2d, which more effectively addresses the majority of challenges related to IoT applications, illustrates a five-layered architecture. The application's service is categorized inside the business layer of this architecture. This layer is tasked with presenting extracted information in a manner suitable for the specific use case. Figure 1.2b, d represent the most renowned architectural among the four. The philosophy underlying a reference architecture can be succinctly summarized as follows: "An IoT architecture must possess comprehensive perception and the capability to make effective decisions by extracting reliable information through the intelligent processing of data acquired via a dependable transmission."

1.2 Evolution of IoT

The vision of IoT has gone under changes since 1999 when the idea was conceived. In earlier days, it was limited to connecting equipment's in industries to optimize usage and production and minimize wastage. But today, IoT Vision includes connection of everything or as per Cisco, Internet of Everything. The main reason of discussing vision of IoT is to understand its journey from past to present and to know what will future look like? There are three ways internet will impact or benefit every business.

Communication: It will convey information about state or health of Things (Equipment, Machine, etc.) beforehand, which can be utilized to cut down the losses due to failure of devices.

Control: Through IoT users will be able to control (remotely) and observe its devices, hence improving the productivity.

Cost: by right decision-making, money loss would be avoided e.g. automatic suggestions to take a device out of service if it is about to fail. In retail chains, advice regarding selling of items [4] which are about expire than which have a later expiry date.

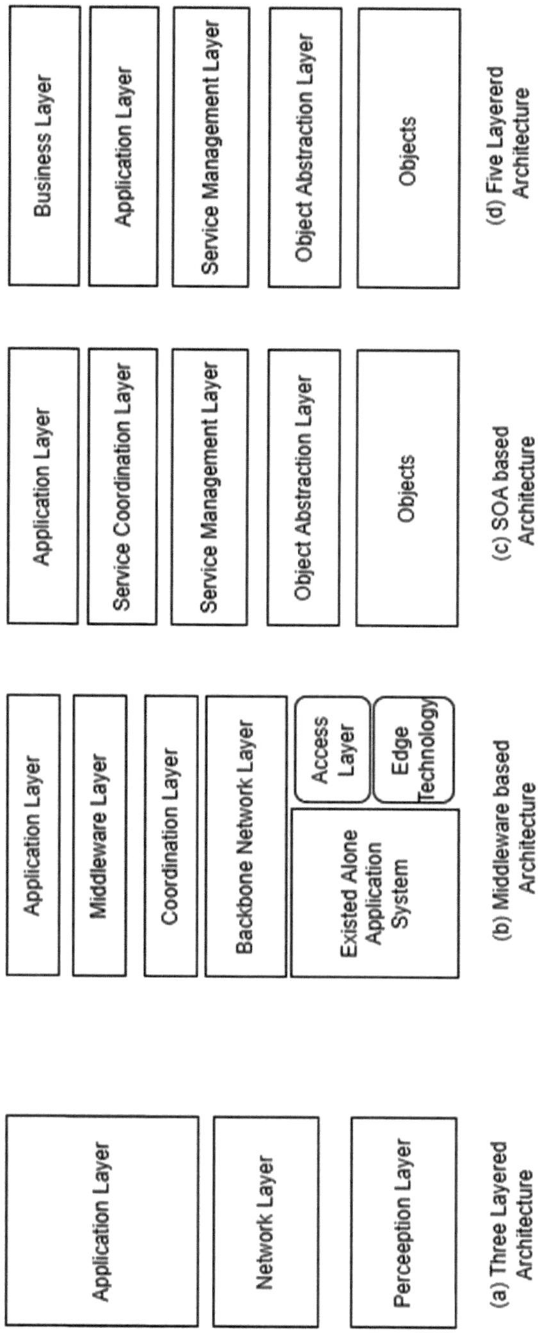

Fig. 1.2 Four distinct IoT architecture in use [8]

An IoT system is all about providing solution for a business scenario. Its requirements are defined by the stakeholders of the business scenario. These are the objectives that architecture of the IoT tries to achieve. By understanding the aspirations and concerns of a stakeholder an optimum architecture can be designed. Hence vision of IoT tells us in brief why reference model for the IoT architecture is required, and how it is designed? In what kind of applications, it is useful. As vision of EU, FP7 [5] clearly tells that at the communication level and at the service level, solutions should be interoperable.

IoT since its inception has gone through various changes. Currently, we are in the stage which can be termed as third generation of the IoT [1]. The first generation started with the inception of the word 'Internet of Things' by Kevin Aston who used the RFID to reduce wastage and bring down costs for Proctor & Gamble. Based mainly on the RFID and Electronic Product Code (EPC) global IoT first generation was the "Generation of Tagged Things". Main function of RFID was to identify and connect things. With the help of EPC-Global network, users and producers could track whole lifecycle of the product. Towards the end of this generation, developers could utilize the maturity of technology in the field of RFID, Wireless Sensor Networks and Supervisory Control and Data Acquisition (SCADA). They harnessed the advancements in the field of energy harvesting as well. The outcome was Wireless Sensing and Identification Platform (WISP). At the system level, physical entities in the real world have well defined virtual entities (VE) which are connected to IoT services and can be manipulated to get the desired services.

Second generation of IoT was Web of Things. Some parallel developments in the scenario paved the way for this. Engineers were facing problems while integrating EPC Global with their products. In the same time period, IPv6 over Low power Wireless Personal Area Networks (6LoWPAN) which was proposed by Internet Engineering Task Force (IETF), started to become popular. Availability of CoAP (Constrained Application Protocol) in the same duration made it possible to convert IoT into web of things by making it easier to integrate day to day objects containing embedded device to web. Now it was possible to search for the tagged objects over web and their status can be read using HTTP. It also cleared the path for third generation of IoT.

Third generation of IoT is marked by intelligent objects, cloud computing. In simple terms, it can be termed as the Social Internet of Things (SIoT). In this scenario, intelligent objects create social relationships and networks independent of their owners, keeping the needs of their master in consideration. Now the approach has been changed from host centric to content centric and the things moved on to cloud for the demand of very large/huge IT resources as clouds provide the cost effectiveness and scalability. Another reason for the use of cloud is that IoT devices are highly heterogeneous and the cloud helps in managing this heterogeneity quite well [1]. These shifts also affected the research paradigms (Fig. 1.3).

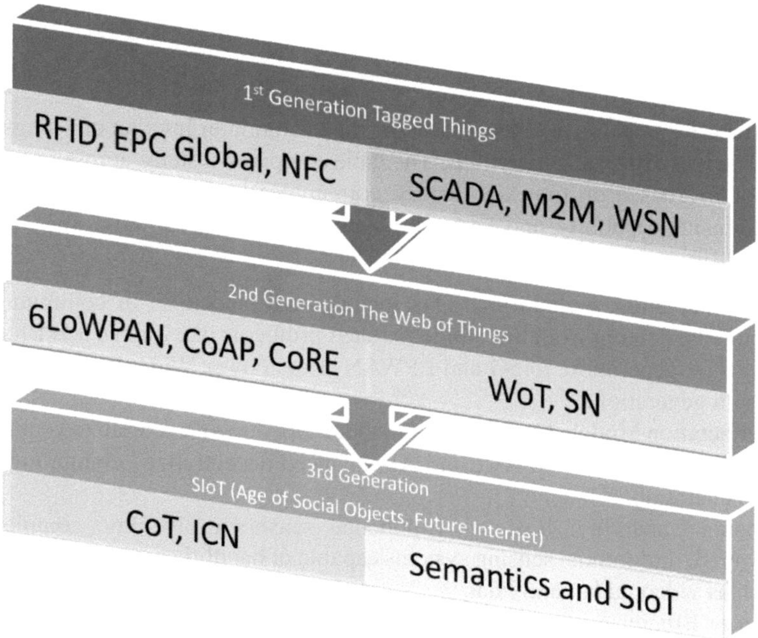

Fig. 1.3 IoT generations with defining technologies [1]

In the first generation the onus was on:

- Improvement in the capabilities of the RFID tag and the hardware that was required to read it.
- Adapting RFID for pervasive computing.
- Improvement of the security and privacy of the RFID (globally unique identification).
- Integration of WSN with RFID.
- Interfacing RFID and WSN with EPC global.

In second generation, research directions also took a turn towards integration of Web.

- Seamless integration of EPC global with Web
- Protocols for the constrained devices.
- Interfacing with the API for the IoT Services

In the third generation, the challenges are manifold and the major research focus is on the:

- Semantics for the data generated (Knowledge extraction)
- Security
- Network Navigability
- Service Discovery

- Complex Mobility Support
- Energy Efficiency
- QoS etc.

Below is a comparative table illustrating the evolution of Internet of Things (IoT) systems across different generations. The table compares systems across four generations of IoT based on four key factors: communication technology, computation model, sensing capability, and energy efficiency.

Key Differentiation Factors:

1. Communication Technology: This includes advancements in communication protocols, ranging from low-range wireless protocols like Zigbee and Bluetooth in the 1st generation, to 5G and LPWAN (Low Power Wide Area Network) in the 4th generation.
2. Computation Model: The computing model evolves from centralized computing in 1st generation IoT to edge and AI-enhanced decentralized computing in the 4th generation.
3. Sensing Capability: Basic environmental sensors evolve into complex AI-powered, multimodal sensing systems capable of handling a wide array of inputs such as video, radar, and lidar.
4. Energy Efficiency: Systems have moved from battery-powered low-efficiency models to ultra-low-power, energy-efficient systems with technologies like energy harvesting.

These references showcase the key milestones in the evolution of IoT, highlighting how the field has evolved in terms of computing, sensing, and communication technologies over the last two decades (Table 1.1).

1.3 Summary

The Internet of Things (IoT) refers to a system of interconnected physical objects that collect, share, and process data to optimize efficiency, reduce waste, and enable intelligent decision-making. Introduced by Kevin Ashton in 1999 using RFID for supply chain management, IoT has evolved significantly due to advances in IC technology and the adoption of IPv6, which allows for the unique identification of billions of devices. Core components of IoT include identification, sensing, communication, computation, semantics, and services. Various architectures—such as IoT-A, RAMI 4.0, IIRA, WoT, and Arrowhead—reflect different needs across industries, with no universal model due to the diversity of applications. Architectures typically follow layered designs incorporating perception, network, middleware, application, and business layers. IoT has developed through four generations: the first focused on RFID and basic sensing; the second introduced web integration using protocols like CoAP and IPv6; the third adopted cloud computing and enabled context-aware, intelligent networking (Social IoT); and the fourth integrates AI and blockchain for

Table 1.1 A detailed comparison between different generations of IoT systems [1, 8]

Generation	System	Communication technology	Computation model	Sensing capability	Energy efficiency
1st Generation (pre-2010)	Wireless Sensor Networks (WSN)-based IoT	Zigbee, Bluetooth, Wi-Fi (Low bandwidth, short range)	Centralized computing (data processed on a central server)	Basic environmental sensors (temperature, humidity, etc.)	Low to moderate efficiency, often battery powered with limited life
2nd Generation (2010–2015)	Cloud-based IoT systems	LTE, Wi-Fi (improved coverage, higher bandwidth)	Cloud computing (data sent to cloud for processing)	Advanced environmental sensing (PM2.5, CO_2, gas sensors)	Energy-efficient communication modules, longer battery life
3rd Generation (2015–2020)	Edge computing-based IoT systems	4G/LTE, LoRa, NB-IoT (low power, wide area networks)	Edge computing (data processed at the network edge)	Context-aware sensing (motion, proximity, cameras, air quality)	High energy efficiency with low-power wide-area networks (LPWANs)
4th Generation (2020–Present)	AI and blockchain-enhanced IoT systems	5G, LPWAN, Wi-Fi 6 (high-speed, low-latency)	AI-based decentralized computing, blockchain for secure data management	Multimodal sensing (AI-enabled cameras, radar, lidar, medical sensors)	Ultra-low power consumption, energy harvesting systems

decentralized, secure, and ultra-efficient systems using 5G and energy-harvesting technologies. These developments have progressively enhanced communication, computation, sensing capabilities, and energy efficiency. Today's IoT spans everything from smart homes to industrial automation, underpinned by cloud, edge, and fog computing. The field continues to evolve, facing research challenges in areas like data semantics, security, mobility, and interoperability. Ultimately, IoT's vision is to connect any device, anytime, anywhere, making it a foundational technology for the digital transformation of society.

References

1. L. Atzori et al., Understanding the Internet of Things: definition, potentials, and societal role of a fast-evolving paradigm. AdHoc Netw. (2017). https://doi.org/10.1016/j.adhoc.2016.12.004
2. S. Chen et al., A vision of IoT: applications challenges, and opportunities with China perspective. IEEE Internet Things J. **1**(4) (2014)
3. D. Evans, The Internet of Things how the next evolution of internet is changing everything. CISCO IBSG (2011)
4. "An introduction to Internet of Things" Part 1 of IoT series by Lopez Research (2013). https://www.cisco.com/c/dam/en_us/solutions/trends/iot/introduction_to_IoT_november.pdf
5. Internet-of-Things Architecture (IoT-A), Project Deliverable D1.2—Initial Architectural Reference Model for IoT. http://www.IoT-a.eu/public/public-documents/d1.2. Accessed 16 Jun 2011
6. E.Y. Nakagawa, P.O. Antonino, F. Schnicke, R. Capilla, T. Kuhn, P. Liggesmeyer, Industry 4.0 reference architectures: state of the art and future trends. Comput. Ind. Eng. **156**, 107241 (2021). ISSN 0360-8352. https://doi.org/10.1016/j.cie.2021.107241
7. M.A. Pisching, M.A.O. Pessoa, F. Junqueira, D.J. dos Santos Filho, P.E. Miyagi, An architecture based on RAMI 4.0 to discover equipment to process operations required by products. Comput. Ind. Eng. **125**, 574–591 (2018). ISSN 0360-8352. https://doi.org/10.1016/j.cie.2017.12.029
8. A. Al-Fuqaha, M. Guizani, M. Mohammadi, M. Aledhari, M. Ayyash, Internet of Things: a survey on enabling technologies, protocols, and applications. IEEE Commun. Surv. Tutor. **17**(4), 2347–2376 (2015). https://doi.org/10.1109/COMST.2015.2444095

Chapter 2
IoT Elements

This chapter explores the fundamental elements and supporting technologies of the Internet of Things (IoT), which together enable the transformation of physical objects into intelligent, connected systems. It presents a deep dive into the components that make IoT functional—ranging from identification and sensing to data communication, processing, semantics, and service delivery. The chapter also emphasizes the importance of enabling technologies like RFID, Wireless Sensor Networks (WSNs), cloud computing, big data analytics, Software Defined Networking (SDN), and Information-Centric Networking (ICN). Through these discussions, the chapter highlights both the capabilities and ongoing challenges in the development and deployment of scalable, secure, and efficient IoT systems.

2.1 IoT Elements

IoT can be understood by its six elements [1] that are utmost necessary to build any IoT system These are depicted in Fig. 2.1 and are as follows:

i. Identification
ii. Sensing
iii. Communication
iv. Computation Services and
v. Semantics

Unique identification of any object or device that constitute IoT system is the basic requirement. This identification helps operate or manage any object throughout world with the help of software. In short, all objects of IoT systems have a unique virtual identity alongside the real ones. Some object might exist in virtual world only. The most important aspect of IoT is to identify an object and to cater it with a matching service. There are two steps of this identification process:

T. K. Sharma et al., *Understanding Internet of Things: Evolution, Enabling Technologies and Challenges*, Studies in Computational Intelligence 1282,
https://doi.org/10.1007/978-3-032-21588-8_2

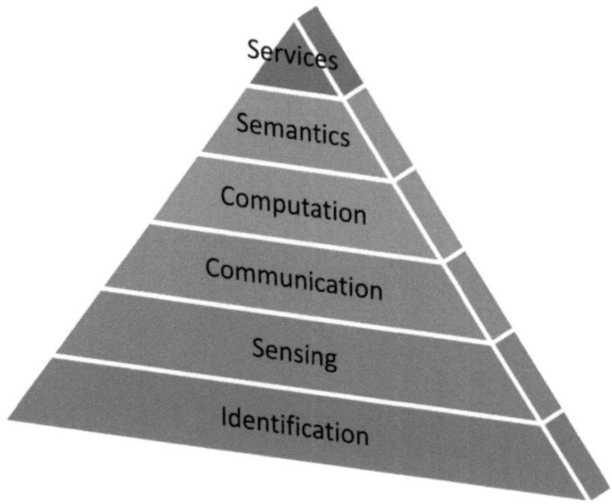

Fig. 2.1 IoT elements

i. Assigning ID to an object
ii. Assigning address to that object

Object ID here stands for "the name given to a sensor" say we have an application for the weather monitoring and we have identified the Humidity Sensor as "H1". Object address is its address within a communication network. In this way to clearly identify an object in a network, we use identification methods. Electronic product codes (EPC) and ubiquitous codes (u-Code) have been used extensively for the identification [2, 3]. As mentioned earlier IPv6 is used for the addressing in IoT along with IPv4. Now a days, a compressed version of IPv6 has been designed by the IETF group, 6LoWPAN [4, 5] is being used extensively in IoT.

In sensing, data is collected from the objects in the network and then passed on to cloud where it is analyzed to help decision making. The ubiquitous wireless sensor networks carry out the work of sensing. A new phenomenon known as Crowd sourcing or crowd-sensing is also used on massive scale now a days [6].

Communication is an important aspect of any IoT system. It is used at every level of IoT. Exchange of information takes place between any two levels or devices. The selection of type of communication technology to be used depends upon various factors such as the terrain, environment, number of nodes connect, distance between nodes and type of data to be delivered, etc. These factors are responsible for the heterogeneity of communication standards used and in turn variety, veracity, and variability of data. The main purpose IoT serve is to deliver smart and specific services to its users. Delivery of these services is also made possible with the help of communication technologies which can be either wired or wireless.

Although a comparatively reliable channel, wired technologies lacks the advantage of mobility. Mobility advantage makes wireless an obvious choice for the communication for different scenarios. Different wireless technologies like Wi-Fi,

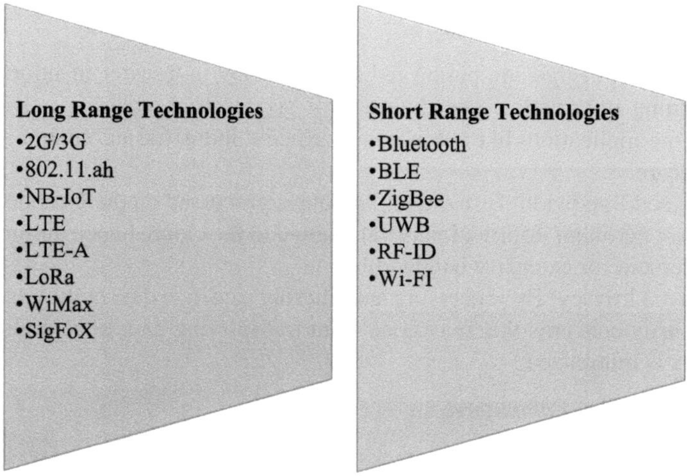

Fig. 2.2 Communication technologies for IoT [5]

Bluetooth, IEEE 802.15.4, LTE-Advanced and in coming future 5G etc. are available for the purpose. Available technologies can be divided into two categories based upon the use case scenario:

- Long Range Technologies
- Short Range Technologies

Figure 2.2 shows long range and short-range technologies being used in IoT. The long-range technologies are 2G, 3G, LTE, LTE-Advanced, WiMAX and 5G in 2020. On the other hand, short-range technologies include Bluetooth and its variants, ZigBee, UWB, and RFID with EPC global.

From designers' perspective, the most important criterion for the selection of a technology is matching of technologies capability with needs of the use case. Many standards have been provided by IEEE communications society. IEEE 802.11, 802.15, 802.16, 802.20, and 802.22 series are the examples [7].

Computation is another element of an IoT system. It serves as the base of intelligent decision making. Every level and device have different amount of computational capability residing with it. It makes the need of a standardized architecture a necessity. Type and quality of services is a direct outcome of the computational abilities of the IoT system. Collection of data by sensors, its conversion into information, extracting the contextual information etc. requires intelligent techniques and different level of computational power [8, 9].

Computation services in the IoT landscape involve the handling and examination of the extensive data produced by IoT devices. These services can be classified into three primary types: edge computing, cloud computing, and fog computing. Each paradigm has its own benefits and difficulties, and the selection among them is based on the unique needs of the IoT application.

Advantages of Edge Computing:

1. Low Latency: Edge computing reduces the delay in transfer of information by computing information near the point of origin. This makes it well-suited for real-time applications like autonomous vehicles, industrial automation, and smart healthcare.
2. Decreased Bandwidth Utilization: As data is processed on the local device, only the most pertinent information is transmitted to the cloud, hence diminishing the requirement for extensive bandwidth.
3. Improved Privacy: By processing and filtering sensitive data at the edge, the risk of security concerns that may arise from transporting data to centralized cloud servers is minimized.

Challenges of Edge Computing:

1. Restricted Computational Resources: Edge devices frequently possess restricted processing capabilities and storage capacity in comparison to centralized cloud data centers, which can pose limitations for intricate data processing jobs.
2. Security: Edge computing improves privacy by confining data within a local network, but it also presents additional safety obstacles, such as protecting scattered devices from potential assaults.
3. Management Complexity: The task of managing and coordinating a significant number of edge devices can be intricate, particularly when it comes to software upgrades, monitoring, and maintenance.

All the Big Data analytics allows the conversion of available information into knowledge that is required for the optimum utilization of resources. Frieder Ganz et al. [10] have rightly suggested that data cannot be processed using a specific technique of information processing like machine learning etc. due to inherent heterogeneity and dynamicity. They have also discussed the proper process of Information processing for the raw data which includes [11]:

(a) Preprocessing (signal or mathematical processing)
(b) Dimensionality reduction using wavelets, Piecewise Aggregation Approximation (PAA/SAX)
(c) Feature Extraction using Machine learning, Complex Event Processing or Logical Inference, etc.
(d) Abstraction or Inference with the same above-mentioned techniques
(e) Semantic Representation

Semantics are required to utilize the knowledge generated for decision making. Semantics are responsible for understanding the demand of user and allocating the appropriate resources on that demand. Semantic interoperability is the key for the utilization of same knowledge for different services [12, 13]. Data alone is not useful until it is converted into knowledge. Knowledge is necessary for right and efficient decision making in the IoT. Data recognition, data analytics, information modeling, resource discovery and utilization are different components of knowledge extraction.

Understanding the demand of the user and allocation of appropriate resource as per demand is done by semantic.

Final part of an IoT system is service. IoT applications are also termed as use cases. These applications can be categorized in to the following type of services [1]:

(a) Identity-related Services,
(b) Information Aggregation Services,
(c) Collaborative-Aware Services and
(d) Ubiquitous Services.

Each use case has its own set of requirements and its own outcomes. But basic requirement, as well as the outcome of every IoT use case, is identification of the objects which helps create the virtual entity of these physical objects. IoT systems also perform collection of information regarding a process or parameters with the help of sensors. Once this data or measurements are taken/collected and forwarded, next level of service is provided by taking decisions based on this data. Last category of services is one in which these decision-making services are provided to anybody, anywhere, anytime.

2.2 Enabling Technologies

An IoT system is amalgamation of a number of technologies. As described in the previous section, the first and foremost requirement of the system is to identify the devices uniquely. After identifying a device in a network and providing it a unique identity, next step is to collect data from it.

2.2.1 RFID, EPC global and WSN

RFID is the most basic technology that enables IoT. IoT started its journey with RFID. RFID along with the EPC global is responsible for the two works:

(a) Identifying the Thing with which it has been attached and this identity is unique throughout the world. It is cheapest mode of Identification.
(b) Connecting things and exchanging the information related with goods and services related to it.

RFID is basically of two types, Active and Passive. EPC global also helps in unique identification of things. EPC global has created various tags for identification. Out of these tags, Class-1 Tag generates 96-bit tag [3] which can address $50 * 10^{15}$ items uniquely.

Near Field Communications is an important technology which has very small applicable range at maximum 10 cm that is communication between reader and a tag can take place maximum upto 10 cm. It works at HF band at frequency 13.56 MHz;

424 kbps is the maximum data rate that is supported by NFC. RFID tag along with EPC global utilizes NFC for identification purposes [2].

After RF-ID Wireless sensor network was the next technology that has helped in the growth of the IoT. Wireless Sensor Network is a mature field now but based on the demands of IoT it also needs some modifications. Given the broad range of user case environment, which shapes the IoT application, every application requires its own set of sensors with different specifications. Apart from sensor networks, a new paradigm has come up for the sensing. It is called Crowd Sensing [14, 15], where mobile phones participate in the sensing and helps in decision making.

The future challenges and research opportunities related with the above fields are following.

Reconfigurable sensing devices are required that can adjust to the scenario and demand of the Use Case. Muhammad S. Khan [3] have designed a reconfigurable RFID tag that can act as generic sensing platform as well as sensing node of the future IoT. It can work in offline as well as online mode as data logging device or Transmitter-Receiver respectively. It helps in sensing data, exchanging information and Routing as well.

One more problem that is associated with the AdHoc Sensor Networks is that they are considered unreliable networks characterized by limited computational power, device mobility, low-capacity channels, etc. A big challenge is to maintain the reliability of communications over such channels. Energy harvesting capability to improve communication can serve good for the scenario. Another way can be to develop routing protocols for this type of scenario by incorporating the rapid recovery mechanism. One such attempt is done by Jiazi Yi et al. [16] who have extended the Depth First Forwarding (DFF) to DFF+ by improving search ordering and used it in different routing protocols, RPL, OLSRv2, and LOADng, to successfully deliver data.

As number of sensors will proliferate in ISM band, spectrum sharing amongst them will become challenge. It will affect the equality in channel access and Spectrum Usage. Feilong et al. [17] have proposed a set of rules called local equilibrium guided autonomous channel switching (LEQ-AutoCS). It helps in equalizing the channel occupations by sensors without generating overhead within its spectrum sensing range. It also reduces spectrum access delay to half of CSMA based systems and 1/50th of TDMA based systems. It is also an effective method for scaling for large industrial wireless application with limited spectrum resources.

Energy efficiency and prolonged life time of WSN is another concern for the researchers. It is mainly done by discontinuous transmissions. But new algorithms are also required to address such issues in case of IoT. One such algorithm based upon the balanced clusters instead of unequal sizes is proposed by Vipin Pal and Girdhari Singh [18] where they have put two thresholds, one on the number of participating nodes in cluster formation and another on the distance of nodes that remained unclustered. Utilizing these two thresholds, they have improved network life time and reduced node death rate.

Improving the performance of sensors in terms of their sensitivity, measurement range, and power consumption is also a research area that can help the future of IoT as it would help provide a better base for effective decision making.

2.2.2 Crowd-Sourcing

Social IoT is a new area of research which focuses upon the identification and utilization of the qualitative and behavioral values of robotic things. Crowd sourcing is a novel concept that has arisen out of the Social IoT. Its conceptual representation is depicted in Fig. 2.3. In this process, different objects mainly mobile phones and smart devices, which are already equipped with various sensors, shares their information for a common cause to help in reliable and efficient decision making. There can be incentive for this service or it may be solely One can understand the Crowd-Sourcing/ Crowd-Sensing as a process of distributed sampling spread over a very large number of individuals. But main issue that comes out of this approach is how to make these Individuals participate in the process. Luis G. Jaimes et al. [6] have discussed many approaches to motivate individuals to participate in the Crowd Sensing to provide a high- quality samples. Analyzing the voice quality is a tough task but Ye Ouyang et al. have developed a wireless analytic tool: Crowd Mi [14] it uses the Crowd Sourcing to model the voice quality in Mobile scenario by mining network indicators from the participants. Another issue related to crowd sensing is the credible interaction between mobile users and based upon these credible interactions, how should one assign the sensing task for crowd-sourcing? Jian An et al. [15] have proposed a model and a hierarchical algorithm based on AHP for crowd sourcing assignment. The model is based on crowd computing and it considers sociality, complexity, and mobility of the user into account.

2.2.3 Machine to Machine Communication

Machine to Machine (M2M) communication is a wired or wireless communication between machines that enables automation possible without human intervention. It is of two types: Capillary M2M and Cellular M2M which transmits in small packet size. It requires high uplink capacity as compared to downlink. A capillary M2M is a highly dense unplanned deployment with poor energy efficiency and high packet loss ratio. Low power link layer technologies are another the main highlights of a capillary M2M network. It uses short range communication technologies for the connectivity such as Zigbee, Wi-Fi, Bluetooth, etc. For wider area, a gateway can be utilized. Low power and lossy network (LLNs) is another name given to this type of networks. If we enable machines to pass information over a cellular network by embedding it with a SIM card it will become Cellular M2M network [19–24].

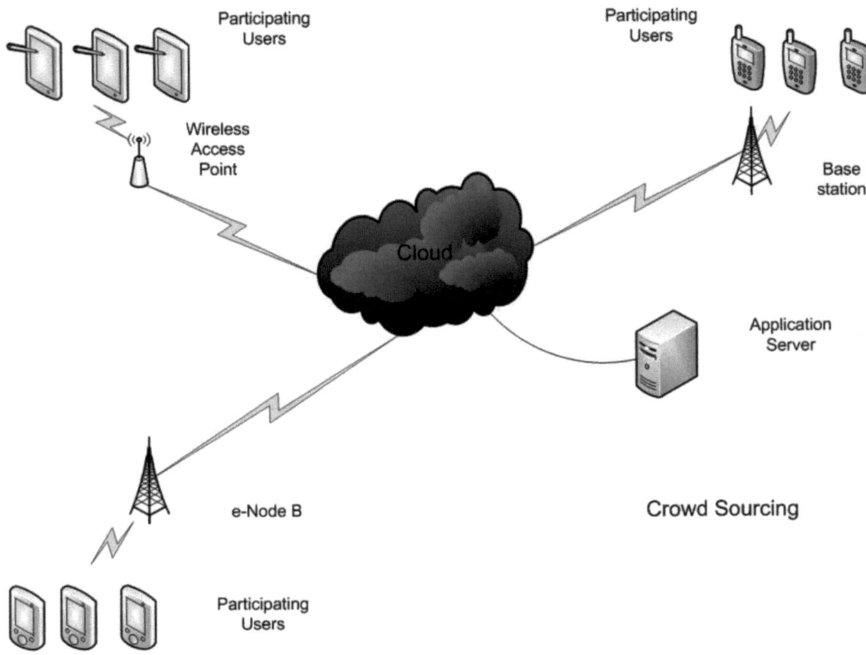

Fig. 2.3 Crowd sourcing

Spectrum is a valuable commodity hence its efficient utilization on massive scale requires intelligent handling in term of spectrum utilization, interference handling, etc. M2M communication also has similar requirement. Cognitive Radio is the technology which can address all such issues. Hence, in the future Cognitive M2M can open latest window of opportunities in this field. M2M traffic has two qualities: first, it has low rate. Second, it has small packet size with correlated transmissions [22]. M2M communication in IoT has many challenges. The number of machines communicating will be numerous and the data would be massive. With so many machines transmitting there will be problems of load control and traffic congestion. Time synchronicity is another challenge associated with M2M communication in IoT. For the identification of machines and other communication requirements overhead may increase. As per Guan Yu Lin [19], contention for the random-access channel [RACH] will increase because of higher number of transmission and it will also increase load on the system. To manage this load efficiently authors have proposed a scheme for the load control of random access. This scheme is based upon an auction model which is used to maximize the valuation sum while using truthful information. This information includes report of low valuation of transmission and estimation of number of devices that are in contention to connect. Reliability of communication needs to be improved in M2M communication. As the nature of wireless channels is dynamic and phenomenon like fading etc. always make reliable communication a challenge. So, new channel coding techniques for M2M scenario are necessary. Security is of utmost importance in M2M scenario just like any other case but here much more is at stake, so authorized access is needed. But performing authentication for each MTC device can create issues of signal congestion and higher consumption of bandwidth in the networks. Jinguo Li et al. [20] have implemented a distributed and group authentication procedure with session key establishment while using asynchronous secret share scheme and Diffie-Hellman key exchange scheme for implementation in LTE-A networks achieving lower bandwidth consumption. Ahasanun Nessa et al. [21] have proposed a set of Joint Network and Fountain Coding (JNFC) schemes to improve the reliability in transmission for M2M communication in LTE-A scenario. Here, they have also addressed the issues of heterogeneity of M2M networks and used clustering and relaying for improving reliability of communication. One of the main issues is whether cellular M2M/MTC will affect the Human Type Communication (HTC) or not? What and how much would be the impact upon the Quality of Service (QoS)? What can be the possible solutions for improving or maintaining the QoS of HTC while maintaining or supporting the MTC to a reliable level as these belong to critical business application? Tiago P. C. de Andade et al. [22] have shown that MTC does affect the QoS provisioning of the HTC strongly for real time traffic. They have proposed 3 PDDCH scheduling algorithms for LTE network to achieve the resource allocation between the two type of communication. They have improved QoS. As far as future of MTC communication in IoT is concerned, it will have to look out for similar kind of solution in its new enabler 5G. Xu Li et al. [23] have given a two phase method to improve the M2M and background traffic. In the approach, packets from machines are aggregated at a virtual gateway before forwarding to a sink. This two phase traffic control has been proposed for a 5G scenario. Another possible direction in future M2M communication is addition of cognitive features to its nodes [24]. The key research area for this would be development of lightweight standardized protocols. It is a good option considering the heterogeniety of devices in IoT. SDN and Cognitive nodes can help in scaling the network size, improving the overall reliability of the network, providing energy efficiency and reducing delays in the network. Naveen M. B. et al. [25] have proposed a modified DRX mechanism to minimize energy consumption in MTC devices. This mechanism incorporates the quick sleep indication for low complexity and low mobility user devices. They have achieved 45% improvement in the energy efficiency and 66% reduction in the computational complexity in the normal coverage scenario while the same results in 63% improvement and 68% reduction for the scenario in enhanced coverage in LTE and LTE-A networks.

2.2.4 Big Data

IoT uses sensors and smart devices for measurements. These sensors collect data digitize them and send over to Internet by different technologies. This transmission creates huge amount of data in the network. As per Dave Evans [26], if a cow is monitored for its health and movements through an IoT application, on average each cow generates 200 mega-bytes of information every year. A connected car generates tens of mega-bytes of data per second. Both examples underline the requirement of big data analytics because data is useful only when the knowledge is extracted out of it.

Big Data is a trend in data mining. Here, one extracts the useful information (knowledge) from huge amount of data with the help of artificial intelligence techniques. It maintains the robustness in results without sacrificing the computational efficiency. Big data does extraction, transformation, and load [ETL] for large data sets. Size of big data is always evolving, it started with tera-bytes and now it may reach zetta-bytes in near future.

Different organizations characterize big data differently. Generally, it is described by 3V's; volume, velocity, and variety. But IBM characterizes it by 4V's; volume for the scale of data, velocity for the analysis of streaming data, variety for different forms of data; veracity for uncertainty. Microsoft characterizes 6V's: volume, velocity and variety are same as before but veracity stands for the trustworthiness of data, variability for complexity of data sets (no. of variables in data sets) and finally visibility signifies one had to have full picture of the data in order to make informative decision [27].

Big data analytics mainly works in three domains (Fig. 2.4):

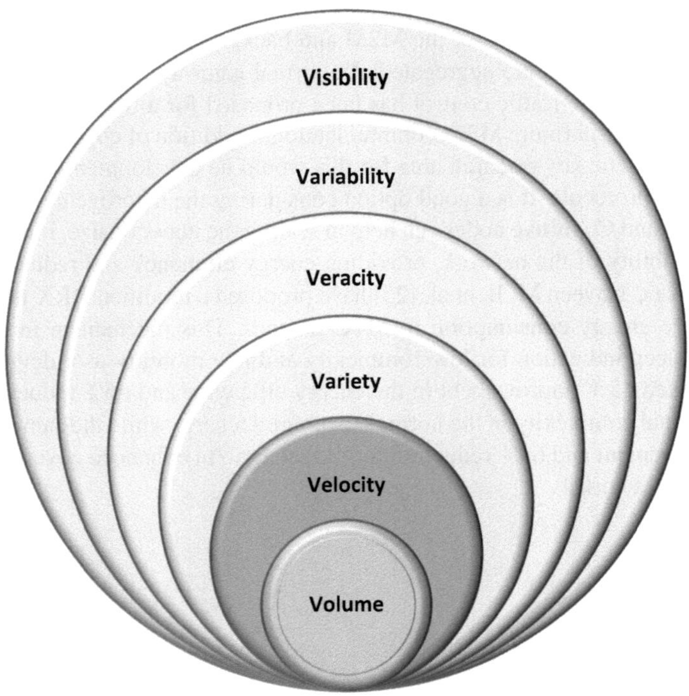

Fig. 2.4 Different characteristics of Big Data

(a) Data domain (searching for patterns) (looks out for velocity, variety, and volume)
(b) Business intelligence domain (making prediction) (visibility, verdict, and value)
(c) Statistical domain (making assumption) (variability, veracity, and validity)

Current generation of IoT applications that utilizes big data are smart supply-chain management, smart energy grids, and syndromic surveillance. This technology suffers from lack of mechanisms to integrate the multiple and heterogeneous distributed data. Heterogeneous sources and data are governed by different policies and access control rules. So, big data technology is not able to ensure the privacy preserving processing. Rahul Pipalkar et al. [28] have presented a thorough survey about the Big Data storage system in IoT. It discussed in detail how to acquire, integrate, process and use the data. Some characteristics, as mentioned above, differentiate the data. Based upon that we can define the type of data:

(a) Multi source high heterogeneity data
(b) Huge scale dynamic data
(c) Low level with weak semantics data
(d) Inaccuracy data

Data analytics is a complex process. To analyse data, it has to be processed under different modules. These are as follows:

(a) Data acquisition and integration modules
(b) Data storage modules
(c) Data management module
(d) Data processing module
(e) Data mining module
(f) Application optimization module

Inclusion of Big Data in to IoT architecture is an important issue. It is required to realize an efficient and effective system. Mohsen Marjani et al. [11] propose a new architecture for 'Big IoT Data Analytics'. Application of Big Data to IoT brings many challenges with it. In data analytics context plays an important role in information processing. Any change in context can change the meaning of available data. Fast extraction of knowledge will help in faster decision making. Complex Event Processing (CEP) is an important tool for the knowledge extraction. It makes context aware knowledge extraction. Meng Ma et al. [29] have proved that multi-pattern event processing can prove to be more effective as compared to the single pattern event processing as it improves the throughput by eliminating the duplicate automata states and redundant computations. Parallel Complex Event Processing is another trend that can be of great use in this field. It can help in detecting and reacting to real-time scenarios. It helps in realizing the "distributed intelligence". Ruben Mayer et al. [30] have proposed a pattern sensitive partitioning model for the data streams. They have also introduced a method to dynamically adapt parallelization degree as per changes in workload for limiting the buffering on event detection.

2.2.5 Software Defined Networking (SDN)

Software Defined Wireless Networking (SDN) [31] improves the network programmability by separating the user and data planes. It can solve the problem of poor flexibility in IoT networks and make these networks adaptable. In [32] with a centralized control plane researcher has managed a heterogeneous network. SDN utilizes the concepts called network slicing and subnet isolation to provide extreme customizability and agility in the network. It fulfils the most important requirement of future IoT Networks as it can provide the Quality of Service (QoS) as per use cases. It will also help in the faster introduction of the new services in the market. There are many issues that can be handled by the SDN in an efficient manner. SDN can help manage the massive data in a better way. It can also help in efficient management of the heterogeneous devices. Allocation of resources can be done optimally with the help of SDN. It can also mitigate interference. Latency reduction and interoperation of the heterogeneous devices can be achieved by this technology.

Zaki et al. [33] have proposed spectrum virtualization where they have pooled physical resource blocks (PRBs) from different eNBs to cater various services is a good solution for providing QoS to cater the massive number of customers. SDN [34] can easily provide solution for the problem of interoperability between different wireless technologies and standards. Its main cause is the heterogeneity of the networks. As both front and back haul of the system will face massive traffic volumes SDN with Cloud can address these issues as well [35] by removing processing intelligence and passing it on to the cloud these problems have been addressed. SDN can also be used for the interference mitigation [36] if the SDN controller utilizes the transmission related information and checks whether two nodes are in the interference range of each other and then allocate the nodes with optimal set of resources so that there is minimum interference is in the system. The one problem that is still not been addressed is that SDN cannot achieve very high utilization of bandwidth in the Internet Service Provider (ISP) backbone networks. Hence, optimization of the SDN over ISP backbone networks is a key issue to work upon.

Handover latency is the main cause for packet loss and loss of services. It requires proper attention of the researchers. SDN can be used to reduce it as the separation of control and data paths helps in this issue. Syed M. Raza et al. [36] have utilized the SDN for mobility control and proposed a new Open Flow based Proxy Mobile IPv6 (OF-PMIPv6). With its help, mobile handover latency was reduced by 43% in reactive and 12% in proactive schemes. Improvement in network performance is also important which can be achieved by reduction in packet loss. Tal Mizrahi et al. [37] have proposed a scheme TIME4 which reduces the packet loss without increasing the capacity of communication link. It also does away with the approach of spare capacity. SDN is a good solution for the issues of latency, heterogeneity, interoperability, etc. as it optimizes path selection and minimizes forwarding time at routers. But it may be the best solution in static and quasi static scenario. As under highly dynamic conditions, centralized controller will have to refresh its paths repeatedly. It will lead to degradation in performance.

2.2.6 *Information Centered Networking (ICN)*

Essence of IoT lays in data centric services. It focuses on information dissemination and sharing. Information Centric Networking is a new paradigm where the focus is on dissemination of information rather than location of information. It is not host centric but information centric approach. It all started with Translating Relaying Internet Architecture integrating Active Directories (TRIAD), an architecture presented in 2000 under a Stanford University Project [38]. After that lots of research projects were undertaken to explore this area. DONA, NDN, NetInf, PURSUIT, CCN, SAIL, CONVERGENCE, CORIN, DATACLOUDS are a few [39–52]. In the coming future, ICN will be mainstay for the IoT use cases. ICN has following the inherent characteristics:

(a) A connectionless receiver driven model
(b) Content based naming and security
(c) In-Network Caching
(d) Name based content dissemination and delivery

ICN utilizes hierarchical and flat names of variable lengths which are unique, self-authenticating and enables any cast retrieval. There are many requirements that are inherent in IoT system:

(a) Scalability
(b) Global reachability
(c) Mobility
(d) Communication diversity (Heterogeneity)
(e) Security
(f) Ubiquitous data exchange
(g) Self -organizing capabilities
(h) Semantic interoperability and data management
(i) Energy optimized solution
(j) Localization and tracking abilities.

Recently, information-centric networking (ICN) has emerged as a new paradigm for the future Internet [38–49]. ICN helps in minimizing the gap between cyber and physical world by replacing Internet Protocol (IP) translation mechanism with utilizing names to address content. CCN [39] which uses in-network caching helps reducing in consumption of energy and bandwidth requirement. MobilityFirst (MF) [45] is an architecture which with a few changes can provide mobility, global reachability, communication diversity, and security. It is achieved by treating service as an object, using gateways for address translation between Global User Identification (GUID) and Local User Identification (LUID), and attribute-based encryption. SAIL [46] is a name resolution system which utilizes hierarchical distributed hash tables. In [45], comparison of Mobility First (MF) and NDN has been provided. MF reduces control overhead and provides higher goodput as compared to NDN which has better performance in terms of latency.

As IoT is interconnection of billions of devices, the traditional approach which generates large overhead can prove to be problematic while on the other side new paradigm of ICN reduces overhead and is well suited for the content dissemination in IoT scenario. Hao Yue et al. [47] have proposed a novel architecture Data Clouds which is based on the information centric networking. Its basic building blocks are communities that are well formed logically as well as physically to help disseminate and share the data more efficiently among interested users. User need not know where the data is stored. In Smart Cities, data centric services will exist everywhere.

ICN can help IoT in fast dissemination of information but it has its own issues as well. Network overhead is a big problem in the NDN. As per [48] NDN over WiFi generates significant overhead resulting in the wastage of data. NDN will improve efficiency in data transfer by transferring the cached content from nodes. If employed in IoT, it can improve the efficiency of the communication. But, cached content has the tendency to become obsolete very quickly. Caching of data in the nodes can help in faster content delivery but it can also create latency and make server hit reduction ratio worse if become obselete. New and efficient protocols are required for routing. Producer mobility management is a big challenge in ICN as forwarding plane keeps on interacting with the information producer. ICN is prone to DoS attacks due to cache updation based on recent requests. Attacker can send false requests to create a problem in cache updation. Researchers can work on these challenges to improve ICN.

2.2.7 *Cloud Computing*

As discussed in the Sect. 2.3, integration of cloud computing was the mark of third generation of IoT [1]. Cloud computing is a technology that is marked by following (Fig. 2.5):

(a) 3S:

    3 Service Models [SaaS, PaaS, IaaS]

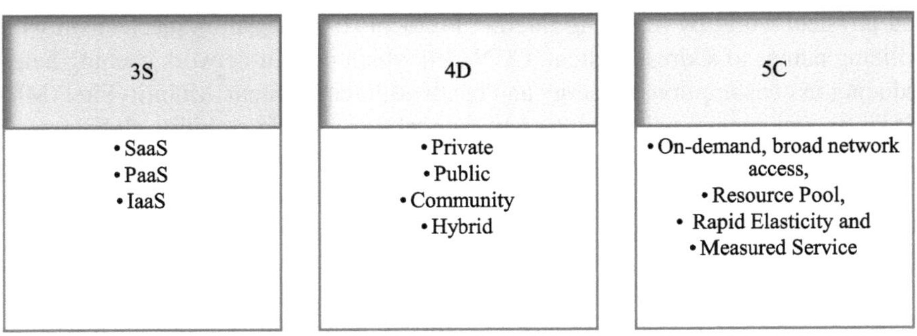

Fig. 2.5 Service model, deployment models and desired characteristics of a cloud

(b) 4D:

4 Deployment Models [Private, Public, Community, and hybrid Cloud]

(c) 5C:

5 Characteristics [On-demand, broad network access, Resource Pool, Rapid Elasticity and Measured Service]

Figure 2.6 shows four important requirements of IoT systems where cloud computing plays important role. The data that will be generated by billions of things in IoT will be humungous. It is far bigger than anything that has come before. As an example, a US smart grid is expected to generate 1000 Peta-bytes of data each year. An autonomous vehicle can generate up to 1 gigabyte of data per second. Now, taking the numbers in to account the storage that is required is huge. This amount of storage is not possible for any user by itself. It can only be provided by the huge data centres. Then, data alone is not useful until it is converted into knowledge. The primary aim of the IoT is provisioning of services on demand. It is not possible without right knowledge and correct decision making.

Cloud has unlimited storage and computation ability that can be provided on demand. Apart from the above two requirements, heterogeneity management and semantic interoperability are other two important needs that can be fulfilled with the help of cloud computing. Cloud normally provides three types of services as

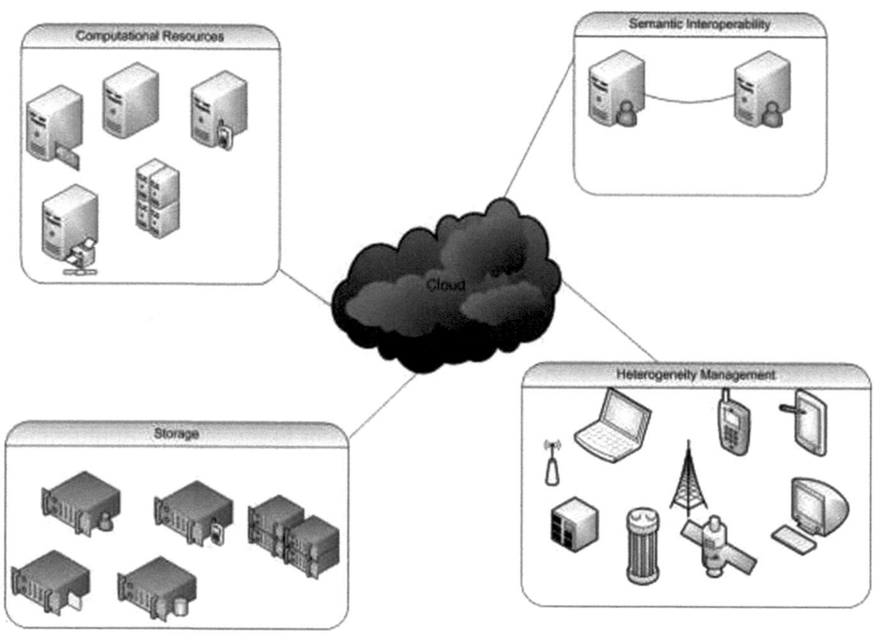

Fig. 2.6 Need for Cloud in IoT

mentioned above and it also helps in five types of works as described in the characteristics part. Cloud can help in resource pooling with rapid elasticity to provide on-demand broad network access. Sherif A. et al. [8] has proposed algorithm for the discovery of sensing resources in a distributed way. It also deploys a virtual sensor network, with the help of virtualization algorithm above a subset of selected IoT devices. It reduces communication overhead utilizes minimum physical resources and has very low complexity. In their work, authors provided this platform to provide sensing as a service. Yi Xu et al. [35] have presented a four layered Cloud Edge Beneath (CEB) architecture to address scalability issues using cloud that would arise in IoT scenario. Salvatore D. et al. [53] have discussed the strategies to enable various type of cloud of things through interaction of cloud with IoT. They have focused on the aggregation and abstraction of heterogeneous resources based on semantics and it can enable things as a service.

2.3 Summary

This chapter identifies six essential elements of IoT systems: identification, sensing, communication, computation, semantics, and services. Each IoT device must have a unique identity, typically enabled by technologies like RFID, EPC, and IPv6/6LoWPAN. Sensing involves data collection via wireless sensor networks or crowdsourcing, feeding into decision-making systems. Communication technologies—ranging from short-range (Bluetooth, ZigBee) to long-range (LTE, 5G)—are selected based on specific use-case needs. Computation is distributed across cloud, edge, and fog computing, enabling real-time analytics and decision-making, though each paradigm presents its own limitations. Semantics allow for meaningful data interpretation and user-demand-driven resource allocation, while services are categorized into identity-based, aggregation, collaborative-aware, and ubiquitous services. The chapter also explores enabling technologies: RFID for identification, WSN for sensing, and cloud computing for scalable data storage and processing. Big Data analytics transforms massive, heterogeneous datasets into actionable insights, despite challenges in privacy and integration. Software Defined Networking (SDN) and Information-Centric Networking (ICN) enhance scalability, latency, and interoperability in complex IoT networks. The chapter concludes by addressing future challenges like reconfigurable sensing, efficient spectrum usage, energy optimization, and machine-to-machine (M2M) communication, highlighting the ongoing need for innovation in protocol design and infrastructure optimization to meet the demands of increasingly dense and diverse IoT ecosystems.

References

1. A. Al-Fuqaha, M. Guizani, M. Mohammadi, M. Aledhari, M. Ayyash, Internet of Things: a survey on enabling technologies, protocols, and applications. IEEE Commun. Surv. Tutor. **17**(4), 2347–2376 (2015). https://doi.org/10.1109/COMST.2015.2444095
2. J. Sung, T. Sanchez-Lopez, D. Kim, The EPC sensor network for RFID and WSN integration infrastructure, in *Proc. of Fifth IEEE International Conference on Pervasive Computing and Communications Workshops (PerCom 2007)* (2007)
3. M.S. Khan, M.S. Islam, H. Deng, Design of a reconfigurable RFID sensing tag as a generic sensing platform toward the future Internet of Things. IEEE Internet Things J. **1**(4), 300–310 (2014)
4. N. Kushalnagar, G. Montenegro, C. Schumacher, *IPv6 Over Low-Power Wireless Personal Area Networks (6LoWPANs): Overview, Assumptions, Problem Statement, and Goals*, vol. 10 (Internet Eng. Task Force (IETF), Fremont, CA, USA, RFC4919, 2007)
5. G. Montenegro, N. Kushalnagar, J. Hui, D. Culler, *Transmission of IPv6 Packets Over IEEE 802.15.4 Networks* (Internet Eng. Task Force (IETF), Fremont, CA, USA, Internet Proposed Std. RFC 4944, 2007)
6. L.G. Jaimes, Idalides, J. Vergara-Laurens, A. Raij, A survey of incentive techniques for mobile crowd sensing. IEEE Internet Things J. **2**(5) (2015)
7. S.S. Dhanda, B. Singh, P. Jindal, Wireless technologies in IoT: research challenges, in *Engineering Vibration, Communication and Information Processing*, ed. by K. Ray, S. Sharan, S. Rawat, S. Jain, S. Srivastava, A. Bandopadhyay. Lecture Notes in Electrical Engineering, vol. 478. Springer, Singapore, (SCOPUS) (2019). Print ISBN: 978-981-13-1641-8
8. S. Abdelwahab, B. Hamdaoui, M. Guizani, T. Zanati, Cloud of things for sensing as a service: architecture, algorithms and use cases. IEEE Internet Things J. **3**(6), 1099 (2016)
9. M. De Donno, K. Tange, N. Dragoni, Foundations and evolution of modern computing paradigms: cloud, IoT, edge, and fog. IEEE Access **7**, 150936–150948 (2019). https://doi.org/10.1109/ACCESS.2019.2947652
10. F. Ganz, D. Puschmann, P. Barnaghi, F. Carrez, A practical evaluation of information processing and abstraction techniques for the Internet of Things. IEEE Internet Things J. **2**(4), 340–354 (2015). https://doi.org/10.1109/JIOT.2015.2411227
11. M. Marjani, F. Nasaruddin, A. Gani, A. Karim, I.A.T. Hashem, A. Siddiqa, I. Yaqoob, Big IoT data analytics: architecture, opportunities, and open research challenges. IEEE Access **5**, 5247–5261 (2017)
12. B. Anuraj, D. Calvaresi, J.-M. Aerts, J.-P. Calbimonte, Dynamic swarm orchestration and semantics in IoT edge devices: a systematic literature review. IEEE Access **12**, 116917–116938 (2024). https://doi.org/10.1109/ACCESS.2024.3446876
13. F.Z. Amara, M. Hemam, M. Djezzar, M. Maimor, Semantic Web and Internet of Things: challenges, applications and perspectives. J. ICT Stand. **10**(2), 261–291 (2022). https://doi.org/10.13052/jicts2245-800X.1029. M. Ganzhaa, M. Paprzyckia, W. Pawowski, P. Szmeja, K. Wasielewska, Semantic interoperability in the Internet of Things; an overview from the INTER-IoT perspective. J. Netw. Comput. Appl. **81**, 111–124 (2017)
14. Y. Ouyang, T. Yan, G. Wang, CrowdMi: scalable and diagnosable mobile voice quality assessment through wireless analytics. IEEE Internet Things J. **2**(4) (2015)
15. J. An, X. Gui, Z. Wang, J. Yang, X. He, A crowdsourcing assignment model based on mobile crowd sourcing in the Internet of Things. IEEE Internet Things J. **2**(5), 358–369 (2015)
16. J. Yi, T. Clausen, U. Herberg, Depth-first forwarding for unreliable network: extensions and applications. IEEE Internet Things J. **2**(3), 199–209 (2015)
17. F. Lin, C. Chen, N. Zhang, X. Guan, X. Shen, Autonomous channel switching: towards efficient spectrum sharing for industrial wireless sensor networks. IEEE Internet Things J. **3**(2), 231–243 (2016)
18. V. Pal, G. Singh, R.P. Yadav, Balanced cluster size solution to extend lifetime of wireless sensor networks. IEEE Internet Things J. **2**(5), 399–401 (2015)

19. G.-Y. Lin, H.-Y. Wei, Auction based random access load control for time-dependent machine-to-machine communications. IEEE Internet Things J. **3**(5) (2016)
20. J. Li, M. Wen, T. Zhang, Group based authentication and key agreement with dynamic policy updating for MTC in LTE-A networks. IEEE Internet Things J. **3**(3), 408–417 (2016)
21. A. Nessa, M. Kadoch, Joint network channel fountain schemes for the machine type communications over LTE-advanced. IEEE Internet Things J. **3**(3), 418–427 (2016)
22. T.P.C. de Andrade, C.A. Astudillo, N.L.S. Doonseca, Allocation of control resources for machine-to machine and human-to-human communications over LTE/LTE-A networks. IEEE Internet Things J. **3**(3), 366–377 (2016)
23. X. Li, J.B. Rao, H. Zhang, Engineering machine-to-machine traffic in 5G. IEEE Internet Things J. **3**(4), 609–618 (2016)
24. A. Aijaz, A. Hamid Aghvami, Cognitive machine to machine communications for Internet of Things: a protocol stack perspective. IEEE Internet Things J. **2**(2), 103–112 (2015)
25. N.M. Balasubramanya, L. Lampe, G. Vos, S. Bennett, DRX with quick sleeping: a novel mechanism for energy efficient IoT using LTE/LTE-A. IEEE Internet Things J. **3**(3), 398–407 (2016)
26. D. Evans, The Internet of Things how the next evolution of internet is changing everything. CISCO IBSG (2011)
27. M. Naeem et al., Trends and future perspective challenges in big data, in *Advances in Intelligent Data Analysis and Applications. Smart Innovation, Systems and Technologies*, vol. 253, ed. by J.S. Pan, V.E. Balas, C.M. Chen (Springer, Singapore, 2022). https://doi.org/10.1007/978-981-16-5036-9_30
28. R.R. Papalkar, P.R. Nerkar, C.A. Dhote, Issues of concern in storage system of Iot based big data, in *2017 International Conference On Information, Communication, Instrumentation And Control (ICICIC)* (2017). https://doi.org/10.1109/ICOMICON.2017.8279126
29. M. Ma, P. Wang, C.-H. Chu, L. Liu, Efficient multi-pattern event processing over high- speed train data streams. IEEE Internet Things J. **2**(4), 295–309 (2015)
30. R. Mayer, B. Koldehofe, K. Rothermel, Predictable low latency event detection with parallel complex event processing. IEEE Internet Things J. **2**(4) (2015)
31. N. Bizanis, F.A. Kuipers, SDN and virtualization solutions for the Internet of Things: a survey. Digital Object Identifier. https://doi.org/10.1109/ACCESS.2016.2607786
32. Y. Zaki, L. Zhao, C. Goerg, A. Timm-Giel, A novel LTE wireless virtualization framework, in *Proc. Int. Conf. Mobile Netw. Manage.* (2010), pp. 245–257
33. F. Granelli et al., Software defined and virtualized wireless access in future wireless networks: scenarios and standards. IEEE Commun. Mag. **53**(6), 26–34 (2015)
34. P. Dely, J. Vestin, A. Kassler, N. Bayer, H. Einsiedler, C. Peylo, CloudMAC—an OpenFlow based architecture for 802.11 MAC layer processing in the cloud, in *Proc. IEEE Globecom Workshops* (2012), pp. 186–191
35. X. Yi, A. Helal, Scalable cloud-sensor architecture for the Internet of Things. IEEE Internet Things J. **3**(3), 285–298 (2016)
36. S.M. Raza, D.S. Kim, D. Shin, H. Choo, Leveraging proxy mobile IPv6 with SDN. J. Commun. Netw. **18**(3), 460–475 (2016). https://doi.org/10.1109/JCN.2016.000061
37. T. Mizrahi, Y. Moses, TIME4: time for SDN. IEEE Trans. Netw. Serv. Manage. **13**(3), 433–446 (2016)
38. D. Cheriton, M. Gritter, TRIAD: A Scalable Deployable NAT Based Internet Architecture. Technical Report (2000). http://www-dsg.stanford.edu/triad/
39. M.F. Bari, S. Chowdhury, R. Ahmed, R. Boutaba et al., A survey of naming and routing in information-centric networks. IEEE Comm. Mag. **50**(12), 44–53 (2012)
40. T. Koponen, M. Chawla, B. Chun, A. Ermolinskiy, K. Kim, S. Shenker, I. Stoica, A data-oriented (and beyond) network architecture, in *ACM SIGCOMM Computer Communication Review*, vol. 37 (ACM, 2007), pp. 181–192
41. N. Fotiou, P. Nikander, D. Trossen, G.C. Polyzos, Developing information networking further: From psirp to pursuit, in *Broadband Communications, Networks, and Systems* (Springer, 2012), pp. 1–13

42. D. Lagutin, K. Visala, S. Tarkoma, Publish/Subscribe for Internet: Psirp perspective in *Future Internet Assembly*, vol. 84 (2010)
43. C. Dannewitz, Netinf: an information-centric design for the future internet, in *Proc. 3rd GI/ ITG KuVS Workshop on The Future Internet* (2009)
44. B. Ahlgren, M. D'Ambrosio, C. Dannewitz, A. Eriksson, J. Golic, B. Grönvall, D. Horne, A. Lindgren, O. Mämmelä, M. Marchisio et al., Second netinf architecture description, 4WARD EU FP7 Project, Deliverable D-6.2 v2.0 (2010)
45. S. Li, Y. Zhang, D. Raychaudhuri, R. Ravindran, A comparative study of Mobility First and NDN based ICN-IoT architecture, in *10th International Conference on Heterogeneous Networking for Quality, Reliability, Security and Robustness*. https://doi.org/10.1109/QSHINE.2014.6928680
46. T. Edwall, Scalable & Adaptive Internet Solutions (Sail) (2011)
47. H. Yue, L. Guo, R. Li, H. Asaeda, Y. Fang, DataClouds: enabling community-based data-centric services over the Internet of Things. IEEE Internet Things J. **1**(5), 472–482 (2014). https://doi.org/10.1109/JIOT.2014.2353629
48. L. Zhang, D. Estrin, J. Burke, V. Jacobson, J. Thornton, D. Smetters, B. Zhang, G. Tsudik, D. Massey, C. Papadopoulos et al., Named data networking (ndn) project. Technical Report ndn-0001 (PARC, 2010)
49. V. Jacobson, D. Smetters, J. Thornton, M. Plass, N. Briggs, R. Braynard, Networking named content, in *Proceedings of the 5th International Conference on Emerging Networking Experiments and Technologies* (ACM, 2009)
50. B. Ahlgren, C. Dannewitz, C. Imbrenda, D. Kutscher, B. Ohlman, A survey of information-centric networking. Commun. Mag. IEEE **50**(7), 26–36 (2012)
51. G. Xylomenos, C.N. Ververidis, V.A. Siris, N. Fotiou, C. Tsilopoulos, X. Vasilakos, K.V. Katsaros, G.C. Polyzos, A survey of information-centric networking research. Commun. Surv. Tutor. IEEE **16**(2), 1024–1049 (2014)
52. G. Tyson, N. Sastry, I. Rimac, R. Cuevas, A. Mauthe, A survey of mobility in information-centric networks: challenges and research directions, in *Proceedings of the 1st ACM workshop on Emerging Name-Oriented Mobile Networking Design-Architecture, Algorithms, and Applications* (ACM, 2012), pp. 1–6
53. S. Distefano, G. Merlino, A. Puliafito, Enabling the Cloud of Things, in *2012 Sixth International Conference on Innovative Mobile and Internet Services in Ubiquitous Computing* (IEEE, 2012), 978-0-7695-4684-1/12. https://doi.org/10.1109/IMIS.2012.61

Chapter 3
Communication Standards and Protocols

3.1 Wireless Communication Standards

Optimal decision-making is core to IoT but these decisions are based upon the data sensed by the sensors. Communication this data from sensors to the cloud is very important. All traditional communication technologies whether internet or cellular standards like 2G/3G or 4G is that they all have provided more capacity for the downlink traffic whereas flow of data is more from sensors to server in IoT. Numerous short-bursty data sessions requires high uplink capacity. Device capabilities must match the system demands. Many standards are available in the market for the purpose; important ones are discussed here.

In multipath environments that are very dense, UWB [1] is a viable candidate for short-range communications. It sustains high data rates. Other characteristics that make this option attractive are very low power, huge bandwidth, and improved performance under heavy noise. It is being used a lot for near field communication. Its orthogonal codes help it achieve immunity against interference on same frequency in the vicinity. Its low power consumption and small transmission power extends the battery life.

IEEE 802.11 is a collection of wireless local area network (WLAN) communication standards. 802.11a standard was made available in 1999 and it provided a throughput of 54 Mbps upto the distances of 35–140 m. 802.11n which was launched in 2009 utilized MIMO technique and it improved its throughput to 100 Mbps upto a range of 250 m. 802.11ac has a version that was specially launched for the IoT purpose and it is called as 802.11ah. 802.11ad provides a throughput of 6.7 Gbps but the distance reduces to a room's length. Table 3.1 provides the comparisons of these standards.

This new standard 802.11 ah is tailor made for the IoT application. It operates at 5 bandwidth ranges (1–16 MHz) in sub-1 GHz license exempt band. The maximum transmission power is 250 mW. Its data rates vary in the range of 150 kbps to 347 Mbps. To support large scale IoT, it has RAW (Restricted Access Window)

T. K. Sharma et al., *Understanding Internet of Things: Evolution, Enabling Technologies and Challenges*, Studies in Computational Intelligence 1282,
https://doi.org/10.1007/978-3-032-21588-8_3

Table 3.1 Comparison of various 802.11 standards

Standard	Frequency band (GHz)	Operational distance (m)	Throughput (Mbps)
802.11a 1999	5	35–120	54
802.11b 1999	2.4	35–140	11
802.11g 2003	2.4	38–140	54
802.11n 2009	2.4/5	70–250	100
802.11ac 2013	5	35	500–1000
802.11ad 2016	60	1–10	6700

in which only the users registered with an access point can transmit. Synchronization frame also helps achieve this aim. In [2], utilizing the back-off states of stations, authors have proposed a new grouping scheme for RAW mechanism and succeeded in achieving improved saturation throughput and energy efficiency of the system under non-cross SIoT boundary case. In [3], stations have been divided in groups for the time-based channel access and by reducing the number of contenders for the channel access communication efficiency has been improved.

In [4], it has been shown that Wi-Fi white space can be utilized to facilitate the M2M communication as they occur for sufficiently long time. It also directs in the direction of Cognitive communication. A project named "Smart House-Keeper" has been developed by Shunzhou Wang et al. [5], which utilize Wi-Fi for communicating to the outer world. Increasing deployment of more and more smart devices in 2.4 GHz bands is congesting the ISM band. Coexistence of the 802.15.4 with the 802.11 is necessary. In [6], a back-off mechanism with adaptive back-off time has been developed for the coexistence of these two technologies. Another effort in the same direction has been made by the Eugene and Soumaya [6] using Markov chain models for prioritizing the traffic for Zigbee technology.

IEEE 802.16 is a standard for the broadband wireless services. WiMAX stands for Worldwide Interoperability for Microwave Access. It has data rates ranging from 8 to 80 Mb/s for fixed or nomadic access. Mobility support is very limited which is its biggest issue. Although for the stationary users it exhibits very high capacity. A newer version 802.16m has improved data rate of 100 Mb/s for mobile stations. There is very limited use of WiMAX for IoT. In [7], authors have emphasized the role that WiMAX can play in the field of m-Health. In [8], authors have discussed the M2M scenario. They have suggested a framework, to achieve larger coverage, of dynamic spectrum allocation for the Smart Grid application in cognitive radio environment. In [8], authors have explored the possibilities of utilizing three different technologies for the Smart City project and provided a comparative analysis of WiMAX with Wi-Fi and LTE for the IoT. It clearly outlines the limitations of WiMAX as alternative to these two. In [9], a model of Smart Grid has been tested in OpNet using ZigBee, WiMAX and PLC scenario. It clearly states the suitability of WiMAX for long-distance high-speed communication that requires satisfactory QoS. In [10], effect of the interference level on the uplink performance has been discussed. It has been shown that the cell edge throughput is a function of the interference over thermal noise. By fixing/selecting level of IoT we can maximize cell edge performance in uplink.

IEEE 802.15.4 is a standard that has been widely used for the IoT applications. Low-rate wireless personal area networks (LR-WPAN) standards. ZigBee has been built on this standard. With data rates 40–250 Kb/s it provides a low-speed communication at very low cost making it popular choice for the resource constrained devices. The specifications of 802.15.4 standards are available on the IEEE 802.15 working group website (IEEE 802.15, 2014).

2G and 3G are outdated in today's world but these can still be utilized for low speed IoT devices. LTE and LTE-A has been a revolution in the field of communication bringing an all IP network. These technologies also provided high speed broadband data. But the specifications of these two technologies are in complete contrast to IoT requirement which seeks higher capacity in uplink in comparison to downlink. Two new technology or versions of LTE-A have been released for the specific need of IoT named, LTE-M and Narrow Band IoT. There are still issues that need to be addressed. Almudena Diaz-Zayas et al. [11] have identified the challenges that should be overcome. These are addressing, identification of devices, charging regarding subscription of resources, security, triggering of the devices as per requirement under sporadic transmission, low mobility and time-controlled communication. Mads Lauridsen et al. [12] have analyzed the coverage and capacity of both the new versions of LTE and results show that LTE-M provides 99% coverage indoor with 10 dB additional loss. For additional 30-dB loss, the coverage comes down to 80% of devices in case of LTE-M while NB-IoT provides the coverage for the 95% of devices in same scenario. The main reason behind this difference in performance lies in the inherent design of the two technologies. LTE-M has been designed for the maximum coupling loss (MCL) of 156-dB while NB-IoT is designed for 164 dB. In this work, it has also been shown that to carry out 4 Uplink and 4 Downlink transmission and receptions per day with full security procedures LTE-M can support devices up to 80,000 per sector while for the same NB-IoT can support 5000 devices per sector. For simple information transfer, the number of devices increases to 1 Million per sector for LTE-M and 25,000 devices per sector for NB-IoT. Naveen M. B. et al. [13] have proved that the accurate estimation of residual Carrier Frequency Offset (CFO) and its compensation at the Base station (eNB) will result in the 22.5% to 55.2% reduction of the MTC device. Jin Yang et al. [14] have stressed on the fact that LTE is designed for the broadband services. On the other hand, IoT for many applications generally have low rate low power and infrequent short burst data transmission. Hence, to achieve efficient and reliable communication with adequate coverage it is necessary that broadband data services must be differentiated from the IoT applications. This work has also proved that increase in payload size also increases the transmission time of MTC over the control plane as compared to MTC over the data plane. This points out that mechanism to address this problem is an utmost necessity that must be addressed. Mukesh Taneja [15] have addressed the problem of reducing the payload in uplink direction by providing a frame work of coexistence of LTE and LPWA in which a gateway is selected for downlink traffic while e Node-B is selected for uplink traffic.

Bluetooth is a technology that has low power transmission and consumption as well. It is low-cost technology suitable for short range (8–10 m) data transmissions. It is a personal area network with IEEE standard 802.15.1. Operating in the ISM 2.4 GHz band it achieves data rates of 1–24 Mb/s. Bluetooth Low Energy (BLE or Bluetooth Smart) is its newer version. Earlier, in 2010 BLE was merged with Bluetooth standard v4.0. Bluetooth is another popular technology to create the IoT. It helps in creating many systems that utilize the smart-phones as sensor. Apple's iBeacon and Google's Eddystone is an example. As a part of IBM Watson initiative, Adegboyega A. et al. [16] have presented Bea Smart, it is an mobile application based upon BLE beacon technology which provides employees and visitors navigation through the building in more contextualized way. Stefano Alleto et al. [17] have presented an IoT based smart museum that will provide a real interactive cultural experience. It uses localization using BLE. CalorieMap [18] is another IoT application based on BLE which monitors the intensity of the activity it utilizes Received Signal Strength Indicator (RSSI) based mapping algorithm for the purpose. Improvement in the Sensors is another area where one has to work its design can optimize the power consumption and also it can help increase the sensitivity of Sensors for better readings. Jan Prummel et al. [19] have presented a BLE transceiver which is a classic architecture with double frequency VCO. It is fabricated in TMSC 55 nm CMOS. If the tags utilizing BLE transmits a packet per second then it can run for a year on a lithium coin cell battery of 240 mAH. Backscatter communication is also a technology that can help reduce power consumption significantly [20].

Change in the duty cycle of the Bluetooth can also help achieve the optimized Power consumption. Using the data compression scheme, power consumption can also be reduced in Bluetooth transmission and reception. Chaco John Deepu et al. [21] have achieved the 18% and 53% reduction in transmission power using a Bluetooth in lossy and lossless media respectively.

LoRaWAN (Long Range Wide Area Network) is technology that is build upon the physical layer technology named LoRa [22]. Its main feature is long range which helps cover a wide area. The bands of operation are 433/867/915 MHz. It has the capability to reach deep inside the structures [23, 24] Its long range and low power characteristics can be attributed to chirp spread spectrum. It can operate on multiple data rate at a time on multiple frequencies. This technology employs star topology. It defines two layers of security with AES encryption with key exchange for network and application security. It also handles the requirements for the pairing of gateway with device. It also decides about channel and data rate selection along with power of operation. Low data rates of 0.3–50 kb/s help this technology to achieve a very long battery life (up to 10 years). Three classes of devices are defined in this technology.

Class A defined for Sensors, class B defined for the Sensor and Actuators and class C is for the powered devices that require low latency. These classes have been defined based on Transmission and Reception windows defined for the device. It suits the situation where higher throughput with smaller ranges is required.

Benny V. et al. [25] have studied the impact of interference on coverage and capacity of the LPWANs. The study is carried out in the city of Aalborg and it signifies that without interference or in low noise environments LoRaWAN provides 99% coverage indoor (Table 3.2).

In presence of external interference, outdoor coverage reduces to 90–95% and 95th percentile failure rate increases significantly to 50%. A case study carried out on the deployment of LoRaWAN by Pierre Newmann et al. [26] shows that communication undergoes degradation in Basements. Work by Seung Jae Na et al. [27] and Stefano Tomasin et al. [28] shows that there is an inherent vulnerability in the LoRaWAN security mechanism which makes it susceptible to security risks. It is because of the random number used in joining procedure which makes it vulnerable to Replay and Denial of Service Attacks. Seung Jae et al. [27] provides the procedure to mask the request message to counter this problem. Sarra Naoui et al. [28] have also provided use of Reputation System in selection of the neighbor node as helper for heavy cryptography.

Adwait Dongare et al. [29] have utilized LoRa/LoRaWAN technology to create a LPWAN architecture OpenChirp to allow multiple users to access a cost-effective solution for Infrastructure Monitoring. It has also provided the challenges that are need to be addressed to in IoT scenario like localization, scalability, security and privacy and spectrum efficiency. Enabling IPv6 over LoRaWAN is another challenge which will help in the interoperability issue of IoT products as LoRaWAN does not use IP. Patrick Weber et al. [30] have provided solution 6LoRaWAN in their work to it where they have adapted the packet of LoRaWAN suitable for use over IP.

Owned by Ingenu, SigFoX operates at 868 MHz. It transmits only three messages in three different bands. Its capability of demodulating at low SNR of −142 dBm makes it very useful in metropolitan scenario. Base station scans every channel for message and retrieves it. The Modulation scheme used is BPSK. In uplink, transmission of payload is limited to 12 bytes while for downlink it is 8 bytes. 26 bytes are reserved for protocol overhead. Downlink mode is only for acknowledgement of reception. 1 to 140 messages can be sent in a day via SigFoX [2]. Kyle Benson et al. [31] have developed SCALE a safe community awareness and altering using SiGFOX as one of the technologies.

Table 3.2 Device classes in LoRAWAN

Type of device	Description of communication capability
Class A	For sensors, each UL transmission window with 2 DL receiving windows
Class B	For sensors and actuators, extra receive window in addition to Class A random receive windows
Class C	For low powered devices nearly continuously open receive windows closed only when this device is sending UL traffic

3.2 Protocol Stack

The protocol stack is crucial for creating reliable connectivity in IoT applications. Figure 3.1 illustrates the many protocols available for implementation within the proposed architecture. This section excludes a treatment of TCP and UDP protocols, focusing instead on a brief overview of the other protocols.

3.2.1 *Protocols of the Application Layer*

The Constrained Application Protocol (CoAP) is founded on REpresentational State transport (REST), a web transport protocol characterized by a stateless client–server architecture. This protocol is mostly utilized for social networking applications and mobile devices. GET, PUT, POST, and DELETE are the methods utilized for clearing ambiguity in data transmission. It is tailor-made for resource-constrained devices, lossy connections and minimal battery consumption. It has been achieved through modifications to HTTP. Four categories of messages are utilized in CoAP transmission: confirmable, non-confirmable, acknowledgment, and reset. The first

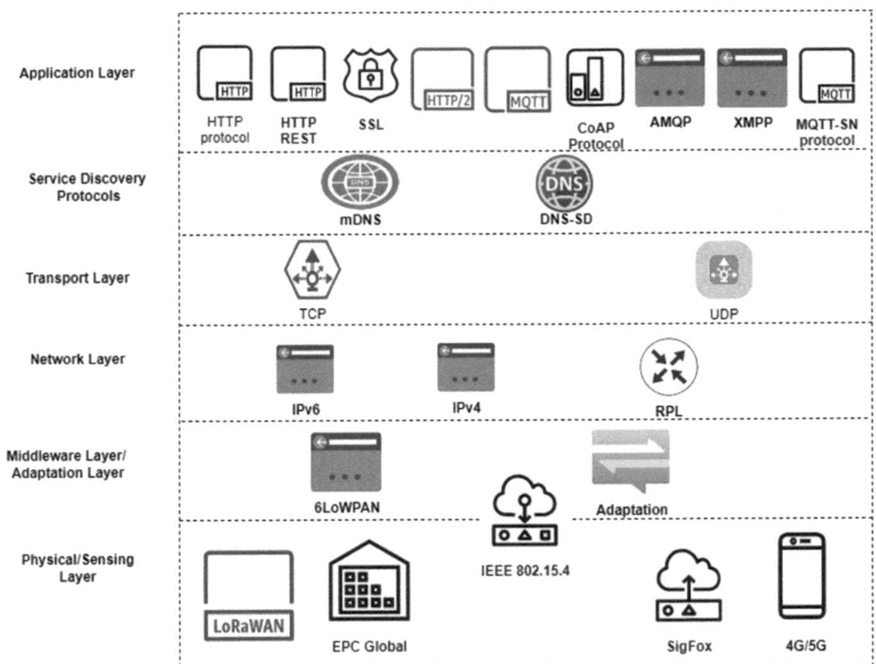

Fig. 3.1 Different protocols working on different layers of the architecture [8]

to enhance reliability and trust in communication. Universal Resource Identifiers (URIs) facilitate service discovery within this protocol.

The Extensible Messaging and Presence Protocol (XMPP) is referenced in IETF standards and was developed by the Jabber open-source community for phone and video communication, as well as multi-party chat. It is a protocol for instant messaging (IM) that is independent of any operating system. It offers highly secure instant messaging that is authenticated and encrypted end-to-end and hop-to-hop. This protocol for IM additionally includes access control and privacy measuring. It employs XML, resulting in significant message overhead.

The Advanced Message Queuing Protocol (AMQP) is engineered for message-oriented environments and necessitates TCP to guarantee reliable communication. It is compatible with publish/subscribe protocols. It accommodates two types of messages: bare messages from senders and annotated messages for receivers. The message layer is constructed above the transport layer.

Message Queue Telemetry Transport (MQTT) is a communications protocol utilized to link middleware and applications with networks and tiny devices. It is constructed on a publish/subscribe architecture and is applicable over unreliable communication channels with little bandwidth to connect constrained devices. It is applicable for structure health assessment, energy monitoring, and intrusion detection. MQTT-SN was designed for application in sensor networks. It offers a broker within the publish/subscribe architecture that facilitates subject indexing and ensures security through the authorization of publishers and subscribers. MQTT has reduced latency relative to CoAP on very lossy connections. However, CoAP is superior on low-loss links. CoAP has superior performance in bandwidth utilization and round-trip latency. CoAP surpasses HTTP in terms of latency and energy efficiency [32].

In IoT, resource discovery is a crucial issue due to the vast scale of the network. DNS Service Discovery (DNS-SD) and Multicast DNS (mDNS) are the two protocols for dynamic and efficient service and resource discovery. mDNS is suitable for small devices as it can operate in case of infrastructure failure without involving additional administration. These attributes render it appropriate for limited devices. In a designated network, DNS-SD protocol allows clients to identify specific services. Multicast addressing and hostnames are resolved by using mDNS and UDP. It also maintains a consistent instance name and ensures a dependable service discovery mechanism [33].

Routing Protocol for Low Power and Lossy Networks (RPL) is a routing protocol for resource-constrained nodes. It is link-independent protocols which adapts IPv6 to enable minimal routing in lossy links. The routers operate in both storing and non-storing modes [34].

6LoWPAN was created to facilitate the transmission of IPv6 datagrams over 802.15.4 networks. It employs four address types to execute compression, forward 802.15.4 packets to the link layer, and identify large packets. 802.15.4 is a standard for low-rate wireless personal area networks (LR-WPAN) that delineates the MAC sublayer and PHY layer. It accommodates data rates ranging from 20 to 250 kbps for both fully functional and reduced functionality devices. A full-function device can act as the coordinator in a Personal Area Network (PAN), whereas reduced-function devices serve as basic nodes inside the PAN. Long Range Wide Area Network (LoRaWAN) is a technique for developing long-range IoT applications. It employs star topology and chirp spread spectrum modulation. It constitutes an open communication standard. SigFox is another communication technique which allows transmission of 1 to 140 messages, of size 12 or 8 bytes, in a day. An overhead of 26 bytes is allowed on both links. EPC Global is one more protocol for facilitating identification, service discovery. It employs five categories of RFID tags for its operations [34].

3.3 Obstacles and Concerns

Security is the paramount component of this architecture. Its scope lies from devices to the cloud and from the cloud to user apps. Most of the communication is wireless making it vulnerable to attackers. Attacker can hack the device or alter the information. It put all of the organization at the risk. Encryption can boost the security in a significant way. Similarly, authentication will ensure legitimate access to organizational data on the cloud and applications. It will be more effective with multi-factor authentication.

3.3.1 Resource Constraints

Limited power, memory, processing capacity and bandwidth are the highlights of IoT devices. It makes lightweight protocols a necessity for operations. Thereby, limiting their ability to utilize strong security processes like error correction, encryption, or complex data formatting. Message Queuing Telemetry Transport (MQTT) is designed for such devices, but its simplicity restricts it functionality in large-scale, secure applications.

3.3.2 Interoperability Issues

Heterogeneity of devices, protocols and standards creates compatibility and integration issues, especially these new devices have to work with legacy systems. It arise due to lack of standardization across industries making seamless communication difficult.

3.3.3 *Scalability*

Numerous devices become part of system that consists of few devices initially. Many protocols struggle to scale due to congestion, packet loss, or inefficient routing. CoAP over UDP may face these problems in dense networks.

3.3.4 *Security and Privacy*

Secure data transmission should be the top priority of protocol design. However, lack of authentication, and access control, weak encryption, susceptibility to spoofing or man-in-the-middle attacks and insecure firmware updates via communication channels are the issues that come with constrained nature of devices.

3.3.5 *Latency and Real-Time Constraints*

Autonomous vehicles, healthcare monitoring, or industrial automation are examples of ultra-low latency applications. Lossy and unreliable networks make it makes situations tougher for real-time communication, especially over or variable networks. TCP-based protocols may induce high latency because of connection setup and acknowledgments procedures.

3.3.6 *Mobility and Dynamic Topology*

Drones, vehicles, or wearables are highly mobile, constituting a dynamic network. It brings continuous changes to network topology. Mostly, communication protocols are not designed for these types of environments. Therefore, in such situations, most important requirement is to maintain session continuity and reliable routing.

3.3.7 *Energy Efficiency*

IoT devices are mostly deployed in remote areas. These devices have limited battery power. Thus, communication protocols are designed for minimizing transmission frequency, packet size, and active time. It conserves energy, makes battery last long and prolongs sensors life. Most of protocols involve frequent handshakes or acknowledgments (e.g., HTTP) for reliable connections, proves themselves to be power-hungry for IoT sensors.

3.3.8 Bandwidth Constraints

IoT networks have very limited bandwidth for operations, especially in low power wide area networks (LPWANs). Therefore, efficient data compression and transmission protocols are required. It mandates for limited-overhead.

3.3.9 Fragmentation and Lack of Unified Standards

The absence of a single unified protocol stack for all IoT applications creates a fragmented landscape of overlapping and sometimes incompatible technologies. It results in increased development time, higher integration cost, and more potential for communication breakdowns.

3.3.10 Protocol Selection and Application Fit

Selection of a protocols for different use-cases is based on different trade-offs in terms of latency, reliability, power consumption, and data rates. It is important to choose the right protocol for the right application. Like, zigbee is suitable for low-power home automation, LoRaWAN would suit the long-range, low-data-rate rural sensing and NB-IoT is a fair choice for cellular-based large-scale deployment.

3.4 Conclusion

Communication protocols are the heart of IoT system design. Numerous challenges are presented by the unique and varied demands of IoT environments. It requires a combination of protocol optimization, adaptive networking techniques, and standardization efforts to solve these issues. 5G, edge computing, and AI-driven network management will play a significant role in addressing these challenges. It also enhances the robustness and scalability of IoT communication.

References

1. A.N. Parks, A.P. Sample, Y. Zhao, J.R. Smith, A wireless sensing platform utilizing ambient RF energy, in *2013 IEEE Topical Conference on Wireless Sensors and Sensor Networks (WiSNet)* (Austin, TX, USA, 2013), pp. 127–129. https://doi.org/10.1109/WiSNet.2013.6488656
2. M. Qutab-ud-din, A. Hazmi, B. Badihi, A. Larmo, J. Torsner, M. Valkama, Performance analysis of IoT-enabling IEEE 802.11ah technology and its RAW mechanism with non-cross slot

boundary holding schemes, in *2015 IEEE 16th International Symposium on A World of Wireless, Mobile and Multimedia Networks (WoWMoM)* (Boston, MA, USA, 2015), pp. 1–6. https://doi.org/10.1109/WoWMoM.2015.7158204
3. L. Zheng, L. Cai, J. Pan, M. Ni, Performance analysis of grouping strategy for dense IEEE 802.11 networks, in *2013 IEEE Global Communications Conference (GLOBECOM)* (Atlanta, GA, 2013), pp. 219–224. https://doi.org/10.1109/GLOCOM.2013.6831074
4. A. Rajandekar, B. Sikdar, On the feasibility of using WiFi white spaces for opportunistic M2M communications. IEEE Wirel. Commun. Lett. **4**(6), 681–684 (2015). https://doi.org/10.1109/LWC.2015.2482973
5. H. Zhao, S. Wang, Y. Liu, X. Hao, Design of an intelligent housekeeping robot based on IoT. J. Inf. Commun. Eng. (JICE) **2**(2), 114–118 (2016), Applied Science and Computer Science Publications, ISSN 2186-9162
6. E.D.N. Ndih, S. Cherkaoui, Adaptive 802.15.4 backoff procedure to survive coexistence with 802.11 in extreme conditions, in *2016 13th IEEE Annual Consumer Communications & Networking Conference (CCNC)* (Las Vegas, NV, USA, 2016), pp. 556–561. https://doi.org/10.1109/CCNC.2016.7444840
7. E.D. Ngangue Ndih, S. Cherkaoui, On enhancing technology coexistence in the IoT era: ZigBee and 802.11 case. IEEE Access **4**, 1835–1844 (2016). https://doi.org/10.1109/ACCESS.2016.2553150
8. R.S.H. Istepanaian, Y.-T. Zhang, Guest editorial introduction to the special section: 4G health—the long-term evolution of m-Health. IEEE Trans. Inf. Technol. Biomed. **16**(1), 1–5 (2012). https://doi.org/10.1109/TITB.2012.2183269
9. Q. Wang et al., Dynamic spectrum allocation under cognitive cell network for M2M applications, in *2012 Conference Record of the Forty Sixth Asilomar Conference on Signals, Systems and Computers (ASILOMAR)* (Pacific Grove, CA, USA, 2012), pp. 596–600. https://doi.org/10.1109/ACSSC.2012.6489076
10. K. Gonchigsumlaa, J. Jigjidsuren, E. Dorj, E. Erdenebileg, The compared analysis of wireless networks for developing Internet of Things, in *2015 2nd International Conference on Information Science and Security (ICISS)* (Seoul, Korea (South), 2015), pp. 1–3. https://doi.org/10.1109/ICISSEC.2015.7371035
11. J.-H. Huh, K. Seo, Smart grid framework test bed using OPNET and power line communication, in *2016 Joint 8th International Conference on Soft Computing and Intelligent Systems (SCIS) and 17th International Symposium on Advanced Intelligent Systems (ISIS)* (Sapporo, Japan, 2016), pp. 736–742. https://doi.org/10.1109/SCIS-ISIS.2016.0161
12. J. Park, S. Jeong, J. Cho, H. Choi, W. Roh, Z. Pi, Interference Level control in mobile WiMAX uplink system, in *2009 IEEE Mobile WiMAX Symposium* (Napa Valley, CA, USA, 2009), pp. 52–56. https://doi.org/10.1109/MWS.2009.32
13. A. Díaz-Zayas, C.A. García-Pérez, Á.M. Recio-Pérez, P. Merino, 3GPP standards to deliver LTE connectivity for IoT, in *2016 IEEE First International Conference on Internet-of-Things Design and Implementation (IoTDI)* (Berlin, Germany, 2016), pp. 283–288. https://doi.org/10.1109/IoTDI.2015.26
14. M. Lauridsen, I.Z. Kovacs, P. Mogensen, M. Sorensen, S. Holst, Coverage and capacity analysis of LTE-M and NB-IoT in a rural area, in *2016 IEEE 84th Vehicular Technology Conference (VTC-Fall)* (2016), pp. 1–5. https://doi.org/10.1109/VTCFall.2016.7880946
15. N. Mysore Balasubramanya, L. Lampe, G. Vos, S. Bennett, Low SNR uplink CFO estimation for energy efficient IoT using LTE. IEEE Access **4**, 3936–3950 (2016). https://doi.org/10.1109/ACCESS.2016.2596679
16. J. Yang, L. Song, A. Koeppe, LTE field performance for IoT applications, in *2016 IEEE 84th Vehicular Technology Conference (VTC-Fall)* (Montreal, QC, Canada, 2016), pp. 1–5. https://doi.org/10.1109/VTCFall.2016.7881080
17. M. Taneja, LTE-LPWA networks for IoT applications, in *2016 International Conference on Information and Communication Technology Convergence (ICTC)* (Jeju, Korea (South), 2016), pp. 396–399. https://doi.org/10.1109/ICTC.2016.7763505

18. A. Akinsiku, D. Jadav, BeaSmart: a beacon enabled smarter workplace, in *NOMS 2016—2016 IEEE/IFIP Network Operations and Management Symposium* (Istanbul, Turkey, 2016), pp. 1269–1272. https://doi.org/10.1109/NOMS.2016.7503001
19. S. Alletto et al., An indoor location-aware system for an IoT-based smart museum. IEEE Internet Things J. **3**(2), 244–253 (2016). https://doi.org/10.1109/JIOT.2015.2506258
20. T.-H. Chiang, Y.-T. Chuang, C.-L. Ke, L.-J. Chen, Y.-C. Tseng, Calorie map: an activity intensity monitoring system based on wireless signals, in *2017 IEEE Wireless Communications and Networking Conference (WCNC)* (San Francisco, CA, USA, 2017), pp. 1–6. https://doi.org/10.1109/WCNC.2017.7925708
21. J. Prummel et al., A 10 mW Bluetooth low-energy transceiver with on-chip matching. IEEE J. Solid-State Circuits **50**(12), 3077–3088 (2015). https://doi.org/10.1109/JSSC.2015.2469674
22. J.F. Ensworth, M.S. Reynolds, BLE-backscatter: ultralow-power IoT nodes compatible with Bluetooth 4.0 Low Energy (BLE) smartphones and tablets. IEEE Trans. Microw. Theory Tech. **65**(9), 3360–3368 (2017). https://doi.org/10.1109/TMTT.2017.2687866
23. C.J. Deepu, C.-H. Heng, Y. Lian, A hybrid data compression scheme for power reduction in wireless sensors for IoT. IEEE Trans. Biomed. Circuits Syst. **11**(2), 245–254 (2017). https://doi.org/10.1109/TBCAS.2016.2591923
24. A. Dongare et al., OpenChirp: a low-power wide-area networking architecture, in *2017 IEEE International Conference on Pervasive Computing and Communications Workshops (PerCom Workshops)* (Kona, HI, USA, 2017), pp. 569–574. https://doi.org/10.1109/PERCOMW.2017.7917625
25. B. Vejlgaard, M. Lauridsen, H. Nguyen, I.Z. Kovacs, P. Mogensen, M. Sorensen, Interference impact on coverage and capacity for low power wide area IoT networks, in *2017 IEEE Wireless Communications and Networking Conference (WCNC)* (San Francisco, CA, USA, 2017), pp. 1–6. https://doi.org/10.1109/WCNC.2017.7925510
26. P. Neumann, J. Montavont, T. Noël, Indoor deployment of low-power wide area networks (LPWAN): a LoRaWAN case study, in *2016 IEEE 12th International Conference on Wireless and Mobile Computing, Networking and Communications (WiMob)* (New York, NY, USA, 2016), pp. 1–8. https://doi.org/10.1109/WiMOB.2016.7763213
27. S.J. Na, D.Y. Hwang, W.S. Shin, K.-H. Kim, Scenario and countermeasure for replay attack using join request messages in LoRaWAN, in *2017 International Conference on Information Networking (ICOIN)* (Da Nang, Vietnam, 2017), pp. 718–720. https://doi.org/10.1109/ICOIN.2017.7899580
28. S. Tomasin, S. Zulian, L. Vangelista, Security analysis of LoRaWAN join procedure for Internet of Things networks, in *2017 IEEE Wireless Communications and Networking Conference Workshops (WCNCW)* (San Francisco, CA, USA, 2017), pp. 1–6. https://doi.org/10.1109/WCNCW.2017.7919091
29. S. Naoui, M.E. Elhdhili, L.A. Saidane, Enhancing the security of the IoT LoraWAN architecture, in *2016 International Conference on Performance Evaluation and Modeling in Wired and Wireless Networks (PEMWN)* (Paris, France, 2016), pp. 1–7. https://doi.org/10.1109/PEMWN.2016.7842904
30. P. Weber, D. Jäckle, D. Rahusen, A. Sikora, IPv6 over LoRaWAN™, in *2016 3rd International Symposium on Wireless Systems within the Conferences on Intelligent Data Acquisition and Advanced Computing Systems (IDAACS-SWS)* (Offenburg, Germany, 2016), pp. 75–79. https://doi.org/10.1109/IDAACS-SWS.2016.7805790
31. K. Benson et al., SCALE: safe community awareness and alerting leveraging the internet of things. IEEE Commun. Mag. **53**(12), 27–34 (2015). https://doi.org/10.1109/MCOM.2015.7355581
32. N. Naik, Choice of effective messaging protocols for IoT systems: MQTT, CoAP, AMQP and HTTP, in *2017 IEEE International Systems Engineering Symposium (ISSE)* (Vienna, Austria, 2017), pp. 1–7. https://doi.org/10.1109/SysEng.2017.8088251
33. A. Siljanovski, A. Sehgal, J. Schönwälder, Service discovery in resource constrained networks using multicast DNS, in *2014 European Conference on Networks and Communications (EuCNC)* (Bologna, Italy, 2014), pp. 1–5. https://doi.org/10.1109/EuCNC.2014.6882683

34. T. Vandervelden, D. Deac, R. Van Glabbeek, R. De Smet, A. Braeken, K. Steenhaut, Evaluation of 6LoWPAN generic header compression in the context of a RPL network. Sensors **24**(1), 73 (2024). https://doi.org/10.3390/s24010073

Chapter 4
Cloud Computing and Its Use in the Internet of Things

4.1 Background

A plethora of opportunities can be created by integrating cloud computing with IoT. Cloud computing can help in creating the scalable and adaptable resources and services that can be available on a click, whenever required. All this can be done by collecting and communicating data through IoT links to cloud. A detailed examination of the advantages and difficulties related to the incorporation of cloud computing into IoT, will be carried out in this chapter.

4.1.1 Comprehending Cloud Computing

Cloud computing provides a wide range of services like virtualized computing resources, such as servers, storage, networking, or development and deployment resources such as middleware, database and tools, or ready to use software applications analytics, and on-demand data storage, data protection and business continuity etc. All these services are scalable, flexible and cost-effective in nature [1].

4.1.2 Cloud Computing Characteristics

- Cloud provides on-demand allocation of computing resources without human involvement.
- Access is not limited to a special type of device but can be accessed over the network by following standard procedures for different type of devices.
- Cloud provides diverse virtual resources to multiple clients in a dynamic manner. i.e. allocation of resources can be adjusted to the changes in demand.

T. K. Sharma et al., *Understanding Internet of Things: Evolution, Enabling Technologies and Challenges*, Studies in Computational Intelligence 1282,
https://doi.org/10.1007/978-3-032-21588-8_4

- Resource utilization is continuously observed, regulated, and documented to ensure complete transparency for service providers and the users.

4.1.3 Cloud Service Models

The cloud-IoT hierarchy is depicted in Fig. 4.1. It shows the different cloud service models and their hierarchy with respect to each other and IoT devices.

- Infrastructure as a Service (IaaS) refers to the provision of virtualized computing resources through the internet, such as AWS EC2 and Google Compute Engine. IaaS (Infrastructure as a Service) denotes infrastructure that is maintained and provided to users on demand, including storage, memory, and networking resources. Management of the infrastructure must permit various forms of consumer connection with the cloud domain.
- Platform as a Service (PaaS) refers to the delivery of hardware and software tools over the internet, specifically designed for application development. Examples of PaaS are Google App Engine and Microsoft Azure. It saves data for testing, setting up, developing, hosting, and maintenance purposes. Upon request, engineers in the software sector proficient in PHP or Java may utilise the software environment, including integrated development environments (IDEs) or application software development stages, such as software development kits (SDKs).
- Software as a Service (SaaS): Provides software applications through the internet, with users paying for a subscription to access them. Examples include Salesforce and Microsoft Office 365.

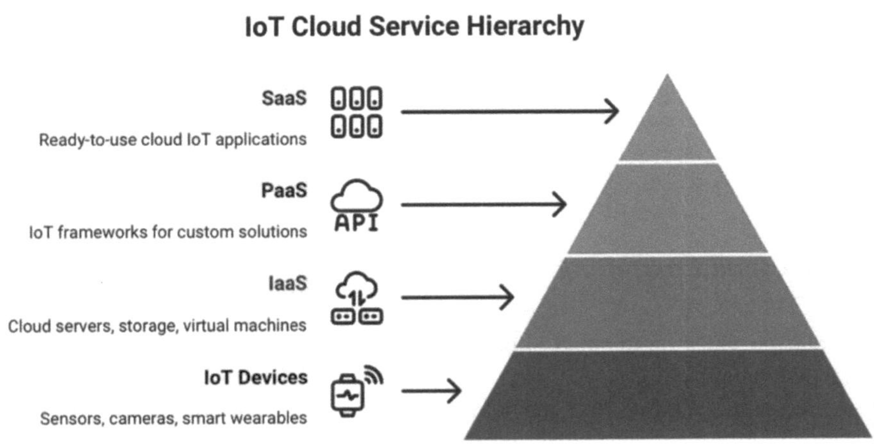

Fig. 4.1 IoT-cloud hierarchy [1]

In addition to the N.I.S.T. definition, four types of clouds can access all services, including Hardware as a Service (HaaS), Database as a Service (DaaS), and Business Process as a Service (BPaas) [2–4, 7]

4.2 Integration of Internet of Things (IoT) with Cloud Computing

The combination of IoT and cloud computing results in a potent synergy that amplifies the capabilities of data collecting, storage, processing, and analysis. It is presented in Fig. 4.2.

4.2.1 IoT Architecture Utilizing Cloud Computing

Multiple layers work in this type of architecture. The bottom layer consists of sensors and actuators. It is called device layer. This layer is responsible for the collection of data and interaction with the environment.

- Second layer which facilitates communication of data from devices to cloud is called gateway layer. It provides managing preliminary data processing which helps in consolidating data for providing services.
- Network Layer ensures data transfer across the networks to cloud data centres.

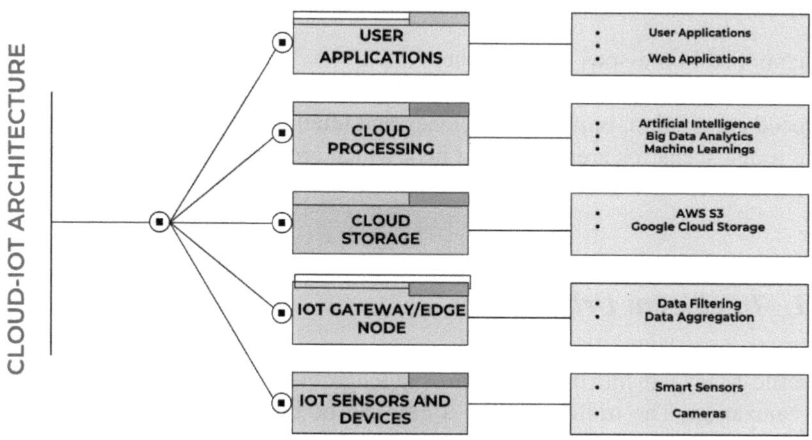

Fig. 4.2 Cloud-IoT architecture [2–6]

- The next layer is cloud layer. It facilitates storage and computational capabilities which are highly scalable. It is also utilized processing, analysis, and management of data.
- Applications and services provide users the refined and processed data. This is accessed through application layer and act as facilitator for the learned decisions.

4.2.2 *Cloud Services for IoT*

Cloud services offer multiple services for the IoT systems. Infrastructure as a service, platform as a service, software as a service, storage as a service, function as a service, networking as a service, database as a service and data-recovery as a service are some of it [7, 8].

- Data Storage and Management: Huge amount of IoT data can be stored and managed by cloud platforms. Amazon S3 and Google Cloud Storage are important example of such services.
- Data Processing and Analytics: AWS Lambda and Google Cloud Dataflow are example of services that provides the option of real-time or batch analytics of IoT data.
- Device Management: AWS IoT Core, Azure IoT Hub offers the option to manage and observe IoT devices.
- Cloud are also used to provide security safeguards and compliance certifications for the IoT data through encryption and access controls.

4.3 Utilizations of Cloud Computing in IoT

As mentioned in previous sections, there are diverse type of applications for which the cloud computing can be integrated with IoT. Its main advantages will result as enhanced automation, improved efficiency and intelligence decision making. Some of the usage scenarios are mentioned here which are enabled by the integration of cloud with IoT.

4.3.1 *Intelligent Urban Areas*

- Traffic Management: It is the main challenge that has come up with increased urbanization. The traffic data for a city can be collected with the help of IoT systems and send to cloud for analysis in real-time. It will help generate alternatives to improve the vehicular-movement, provide more public transportation to decrease the traffic jams [9].

- Enhanced Public Safety: IoT devices such as cameras can provide their feed to cloud through a high-speed link. Clouds offer advanced capabilities to analyse such huge amount of data for surveillance, environmental monitoring, and emergency response systems. They analyse and carry out identification work to raise alarms which enhances urban safety and security [10].

4.3.2 Healthcare

- Remote Patient Monitoring: IoT systems integrated with cloud can be used for real-time monitoring of patients. Large amount of data generated by the wearable gadgets and sensors about a patient. It is also broadcasted to the cloud for the ongoing monitoring and analysis. It facilitates prompt medical treatments [11].
- Telemedicine: Cloud-based combined with IoT devices can also be used for providing video-consultation, online-doctor visits, e-prescriptions etc. The IoT-devices ensure the collection of vitals while cloud provides the interaction platform and patient-record storage. Teladoc, Amwell, Practo etc. are few examples of such online-platforms. These types of platforms are especially useful for third world countries or remote areas [3].

4.3.3 Agriculture

- Precision Farming: IoT sensors helps in collecting the data related to soil conditions which identify nutrients level (N, P, K), moisture content etc. These devices also provide weather patterns, and crop health data. Cloud helps in analytics part and getting the predictive analytics on crop yields and fertilizer and irrigation requirements etc. In this way the yield can be enhanced and wastage can be minimized.
- Livestock Management: Sensors can provide the information on the location, movements, activity levels, nutrition levels and health of animals. At cloud level, all this information can be analysed to find out the patterns that relate to safety of livestock, predict their diseases, optimize their feeding schedules, automation of farm operations. Thereby, enhancing productivity, ensuring safety of livestock [12].

4.3.4 Industrial Internet of Things

- Predictive Maintenance: Industrial equipment is attached with the sensors to collect the operational parameters such as pressure, temperature, vibrations etc. These parameters are transmitted to cloud and performance is calculated with the help of this data. It helps in identifying the anomalies in patterns and predicting

failures. Alarms can be set to prevent equipment breakdowns and reduce downtime by offering timely maintenance. It extends the lifetime as well [6].
- –Supply Chain Optimization: In supply chain operations, most of the optimization can be achieved by tracking goods, inventory, and transportation. This data can be collected with RF-ID tags based IoT systems, QR codes, GPS-enabled IoT tracking system. The data can be collected about a product throughout the world to update the inventory records, manage delivery schedules, predict requirements. Cloud plays the centralized role in receiving this information, process these, and store the details. It helps in bringing the retailers, suppliers and warehouse on the same platform. Thus, provide accurate forecasts, improves visibility, minimizes the delivery delays and enhances transparency [13].

4.3.5 Intelligent Residences

- Home Automation: Smart home devices like smart lights, thermostats, air-conditioners, security cameras, voice assisted-locks etc. all need to collect data and transmit that to cloud for intelligent decision-making. Cloud enables voice assisted alarms, automation rules, and sends real-time alerts for multiple objectives [14].
- Energy Management: Energy saving is the biggest concern for the society. IoT-based systems such as smart meters and appliances gather energy usage data and convey it to cloud. The data is processed and cloud extracts the insights about consumption, wastage, and usage patterns. It then forecast demands, map it to the schedules and which is then evaluated in the cloud to enhance energy utilization and minimize expenses [14].

4.4 Advantages of Cloud Computing in IoT

Integration of cloud computing and IoT offers multiple advantages, cultivate innovation and enhances efficacy.

4.4.1 Scalability

- Scalable Resources: One of the major challenges to IoT systems is scalability. Cloud offers solution offers multiple advantages related to scalability. It not only allows the users to adjust resources as per their requirements but also allow system to adjust resources for adapting to sudden change in data volumes and processing requirements of IoT devices. It also allows system to add and remove IoT devices as and when need arise [15].

- The worldwide infrastructure of cloud services enables the expansion and operation of IoT applications at global scale.

4.4.2 *Cost Effectiveness*

- Cloud computing offers subscription plan where user pays for the package of services of its choice. The most popular model is pay-as-you-go which allows user to be charged for actual computation, storage and bandwidth resources they have utilized [3, 4].
- Cost Efficiency: The architecture of cloud allows to share resources worldwide. In this way, it minimizes the logistics requirements and maintenance costs. Customers organizations can optimize their expenditure by utilizing cloud services [3, 4].

4.4.3 *Improved Data Analytics*

- Enormous data is generated by IoT devices. It is a big challenge to process such a huge amount of data. Sophisticated tools such as AWS Elastic MapReduce (EMR), Azure HDInsight, Apache Spark etc. are utilized for processing the enormous volumes generated by IoT devices. Similarly, data warehousing tools to store and query massive datasets, stream processing tools for real-time analysis, data visualization tools for creating interactive charts, database tools to store structured and unstructured data are provided by cloud. On the top of this cloud provides scalable storage for the variable data requirements that can arise in IoT applications [16].
- Real-Time Insights: These above-mentioned tools help in generating the real-time insights into this IoT data. It means that applications can generate immediate alerts to the variations and anomalies detected in the data. Similarly, machine-settings can be adjusted and the dashboards can be changed dynamically by the use of these tools.

4.4.4 *Enhanced Security*

- As the data collected by the IoT devices is transferred to cloud. It stores critical information and personal data. To ensure security of this sensitive user data, cloud employs various measures. Data is encrypted while it is stored and communicated data so that it remains unreadable for unauthorized user. Multi-factor authentication, secured login, role-based access control are used for the delivery of services for the specific role. Cloud also follows international data protection standards such as HIPAA, GDPR etc. [17].

- Frequent Updates: The attacks, threats and vulnerabilities evolve with time. Therefore, regular security updates and patches are released from time-to-time. These update covers security weaknesses, improved protection methods and about latest security standards [18].

4.5 Difficulties of Cloud Computing in IoT

Despite having multiple advantages, there are multiple difficulties in integration of cloud computing with IoT.

4.5.1 Ensuring the Security and Privacy of Data

- Data Breaches: Hackers keeps updating their methods of data breach. It makes it important to devise methods to avoid unwanted access and breaches in cloud.
- Privacy Concerns: Healthcare systems and smart homes and other such IoT systems gathers user's private information. Its leakage can cost heavily to the person. Therefore, ensuring responsible collection, secure storage, and usage of data for only intended purpose is must. New privacy preserving methods should be developed to ensure all these requirements.

4.5.2 Latency

- Real-Time Processing is required for the latency critical IoT applications. Autonomous vehicles, healthcare monitoring apps and industrial robots are few such applications where instant responses are required. High latency can lead to loss of life, money and infrastructures in these cases. Therefore, inclusion of high-speed routing and networks, utilization of edge computing can be suitable options.

4.5.3 Data Management

- Data Quality: Data gathered by IoT devices does contains errors. Further, during transmission addition of noise affects the precision, coherence, and reliability of data. Data should be free from these inconsistencies only then it can be used for analytics and prediction. Therefore, data should be filtered from errors, duplicates should be removed, inconsistencies, and readings should be validated.
- Data Integration: The inherent heterogeneity of IoT devices reflects into data generated through such systems. Multiple sources, different data formats and

heterogeneous platforms. It requires that the data should be merged, standardized and synchronized to get effective analysis.

4.5.4 Network Connectivity

- Network Reliability: Data is collected from varied terrains by deploying different sensors with limited power. It transmitted wirelessly in real-time to gateway and further. Small network failure can result in the loss of data, delay in delivery or system malfunction. Therefore, stable, continuous, and uninterrupted connection is required for IoT devices.
- Bandwidth Constraints: Bandwidth is very scarce resource. Its optimum utilization is the prime requirement of every communication system. Techniques like data compression, edge computing, and sending only essential data can be employed to achieve the optimization of the transmission of data.

4.6 Outlook for the Future

The future looks bright for the amalgamation of cloud computing and IoT.

4.6.1 Edge Computing

- Decentralized Processing: Edge computing facilitates the data processing near origin. It is utilized in real-time scenarios where latency requirements are very stringent. It not only decreased latency and bandwidth consumption [19].
- Hybrid Architectures: In coming times many applications come up that will have dual requirements. In such cases, a hybrid architecture will be required for stringent latency requirements edge computing will be employed while less stringent cases will be handled by cloud architecture [20].

4.6.2 Artificial Intelligence and Machine Learning

- Advanced Analytics: AI ML algorithms can also be utilized for the creation of intricate data analytics models. These models can predict 'what will happen' and 'what should be done'. It will help in analysing the IoT data to unveil deeper relationships and patterns [21].
- Automated Decision-Making: Use of AI-ML will exclude human from decision-making procedures. This automation will rely on analysis of real-time data and

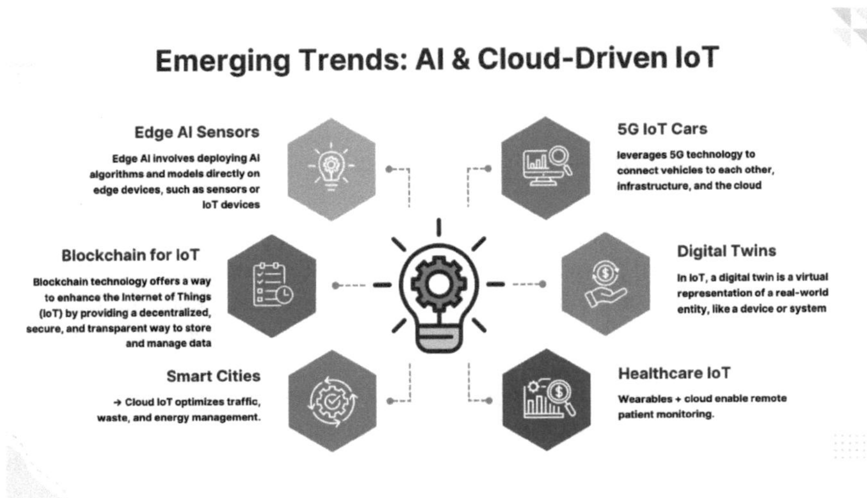

Fig. 4.3 Emerging trends in AI and cloud-driven IoT

will adjust machine settings, send alerts, or activate safety protocols promptly, resulting in enhanced efficiency and promptness [22].

4.6.3 *6G Connectivity*

- Enhanced Data Transmission: Integration of 6G technology is expected to provide 100–1000 times more data-speed. It will improve the capabilities to new levels that will pave the way development of applications for future. Network connectivity will be more reliable resulting is improved data quality for IoT applications [23].
- Large-scale IoT Deployment: The enhanced capacity of data connectivity will allow for the integration of numerous devices (Fig. 4.3).

4.6.4 *Blockchain Technology*

- Data Security: Inclusion of decentralized and tamper-proof blockchain technology will improve the trust as it will make it difficult for attackers to alter data making services more secure [13, 18].
- Transparent Transactions: Blockchain technology will create verifiable communication making it fully full transparent and traceable. It will increase trust, reduce fraud, and improve accountability in cloud-based IoT networks [24].

4.7 Summary

The integration of cloud and IoT has transformed multiple industries by offering scalable, efficient, and intelligent data management and analytics. It has helped in achieving improved decision-making, increased operational efficiency, and greater consumer experiences. Though, there are issues of data security, latency, and connectivity but progress in technology and infrastructure has helped in creating a positive outlook for the future. Upcoming inclusion of AI-ML, 6G and blockchain technology to cloud computing and IoT will result not only in the solution for these challenges but also provide the innovative solutions that will change the face of several industries.

References

1. D. Kapil, P. Tyagi, S. Kumar, V.P. Tamta, Cloud computing: overview and research issues, in *2017 International Conference on Green Informatics (ICGI)* (Fuzhou, China, 2017), pp. 71–76. https://doi.org/10.1109/ICGI.2017.18
2. V.(J.R.) Winkler, Chapter 2—Cloud computing architecture, in *Securing the Cloud, Syngress*, ed. by V.(J.R.) Winkler (2011), pp. 29–53. ISBN 9781597495929. https://doi.org/10.1016/B978-1-59749-592-9.00002-6
3. M.A. Mizani, Chapter 17—Cloud-based computing, in *Key Advances in Clinical Informatics*, ed. by A. Sheikh, K.M. Cresswell, A. Wright, D.W. Bates (Academic Press, 2017), pp. 239–255. ISBN 9780128095232. https://doi.org/10.1016/B978-0-12-809523-2.00017-0
4. D. Talia, P. Trunfio, F. Marozzo, Chapter 2—Introduction to cloud computing, in *Computer Science Reviews and Trends, Data Analysis in the Cloud*, ed. by D. Talia, P. Trunfio, F. Marozzo (Elsevier, 2016), pp. 27–43. ISBN 9780128028810. https://doi.org/10.1016/B978-0-12-802881-0.00002-0
5. A. Al-Dulaimy et al., The computing continuum: from IoT to the cloud. Internet Things **27**, 101272 (2024). ISSN 2542-6605. https://doi.org/10.1016/j.iot.2024.101272
6. D. Mourtzis, 2—Industry 4.0 and smart manufacturing, in *Manufacturing from Industry 4.0 to Industry 5.0*, ed. by D. Mourtzis (Elsevier, 2024), pp. 13–61. ISBN 9780443139246. https://doi.org/10.1016/B978-0-443-13924-6.00002-8
7. S. Azouzi, S.A. Ghannouchi, Z. Brahmi, Study of e-learning system based on cloud computing: a survey, in *Proc. Int. Conf. Intell. Syst. Design Appl.* (Springer, Cham, Switzerland, 2018), pp. 532–544. https://doi.org/10.1007/978-3-030-16660-1_52
8. H. Eljak et al., E-learning-based cloud computing environment: a systematic review, challenges, and opportunities. IEEE Access **12**, 7329–7355 (2024). https://doi.org/10.1109/ACCESS.2023.3339250
9. A. Asha, L. Rajeshkumar, V.S. Pandi, D. Shobana, J. Lakshmi Priya, M. Dayanidhy, Implementing cloud computing with Internet of Things (IoT) technologies: novel approaches to data management and service delivery innovation, in *2024 Global Conference on Communications and Information Technologies (GCCIT)*, (Bangalore, India, 2024), pp. 1–6. https://doi.org/10.1109/GCCIT63234.2024.10862745
10. C. Fontes, E. Hohma, C.C. Corrigan, C. Lütge, AI-powered public surveillance systems: why we (might) need them and how we want them. Technol. Soc. **71**, 102137 (2022). ISSN 0160-791X. https://doi.org/10.1016/j.techsoc.2022.102137
11. S. Iranpak, A. Shahbahrami, H. Shakeri, Remote patient monitoring and classifying using the internet of things platform combined with cloud computing. J Big Data **8**, 120 (2021). https://doi.org/10.1186/s40537-021-00507-w

12. M. Caria, J. Schudrowitz, A. Jukan, N. Kemper, Smart farm computing systems for animal welfare monitoring, in *2017 40th International Convention on Information and Communication Technology, Electronics and Microelectronics (MIPRO)* (Opatija, Croatia, 2017), pp. 152–157. https://doi.org/10.23919/MIPRO.2017.7973408
13. M.D. Karumanchi, J.I. Sheeba, S.P. Devaneyan, Cloud based supply chain management system using blockchain, in *2019 4th International Conference on Electrical, Electronics, Communication, Computer Technologies and Optimization Techniques (ICEECCOT)* (Mysuru, India, 2019), pp. 390–395. https://doi.org/10.1109/ICEECCOT46775.2019.9114692
14. A. Kudryavtsev, V. Zatsepina, I. Pavlov, Implementing cloud computing in smart cities, in *2023 3rd International Conference on Technology Enhanced Learning in Higher Education (TELE)* (Lipetsk, Russian Federation, 2023), pp. 255–260. https://doi.org/10.1109/TELE58910.2023.10184349
15. M. Razian, M. Fathian, H. Wu, A. Akbari, R. Buyya, SAIoT: scalable anomaly-aware services composition in cloud IoT environments. IEEE Internet Things J. **8**(5), 3665–3677 (2021). https://doi.org/10.1109/JIOT.2020.3023938
16. V.K. Prasad et al., Efficient resource utilization in IoT and cloud computing. Information **14**(11), 619 (2023). https://doi.org/10.3390/info14110619
17. B. Alouffi, M. Hasnain, A. Alharbi, W. Alosaimi, H. Alyami, M. Ayaz, A systematic literature review on cloud computing security: threats and mitigation strategies. IEEE Access **9**, 57792–57807 (2021). https://doi.org/10.1109/ACCESS.2021.3073203
18. L. Albshaier, A. Budokhi, A. Aljughaiman, A review of security issues when integrating IoT with cloud computing and blockchain. IEEE Access **12**, 109560–109595 (2024). https://doi.org/10.1109/ACCESS.2024.3435845
19. X. Ren et al., AI-Bazaar: a cloud-edge computing power trading framework for ubiquitous AI services. IEEE Trans. Cloud Comput. **11**(3), 2337–2348 (2023). https://doi.org/10.1109/TCC.2022.3201544
20. S.S. Jazaeri, S. Jabbehdari, P. Asghari et al., Edge computing in SDN-IoT networks: a systematic review of issues, challenges and solutions. Cluster Comput. **24**, 3187–3228 (2021). https://doi.org/10.1007/s10586-021-03311-6
21. N. Mohamed, L. Sridhara Rao, M. Sharma, S.B. Rajasekaran, B.S. Alfurhood, S. Kumar Shukla, In-depth review of integration of AI in cloud computing, in *2023 3rd International Conference on Advance Computing and Innovative Technologies in Engineering (ICACITE)* (Greater Noida, India, 2023), pp. 1431–1434. https://doi.org/10.1109/ICACITE57410.2023.10182738
22. X. Wen, Y. Zheng, The application of artificial intelligence technology in cloud computing environment resources. J. Web Eng. **20**(6), 1853–1866 (2021). https://doi.org/10.13052/jwe1540-9589.2067
23. G. Manogaran, T. Baabdullah, D.B. Rawat, P.M. Shakeel, AI-assisted service virtualization and flow management framework for 6G-enabled cloud-software-defined network-based IoT. IEEE Internet Things J. **9**(16), 14644–14654 (2022). https://doi.org/10.1109/JIOT.2021.3077895
24. Z. Cai et al., RBaaS: a robust blockchain as a service paradigm in cloud-edge collaborative environment. IEEE Access **10**, 35437–35444 (2022). https://doi.org/10.1109/ACCESS.2022.3161744

Chapter 5
Comprehending the Concepts of Fog and Edge Computing

5.1 Definition of Fog Computing

The concept of cloud computing when extended and brought closer to the edge of network it is termed as Fog computing, or fogging. It realizes the decentralized computing framework. This paradigm distributes data, storage, and applications between the data source and the cloud [1, 2]. "*Fog computing is a horizontal architecture that is created for the allocation of resources and services, such as computation, storage, control, and networking, throughout the entire spectrum from the Cloud to devices.*".

- Architecture: Fog nodes can be put in any location that has network connectivity, such as on-premises servers, routers, gateways, or dedicated fog devices.
- Scenarios: There are multiple scenarios where fog computing nodes need to be deployed near the data collection scenarios or points. E.g. in autonomous cars, the latency requirements are of '*milliseconds*' for the decisions such as lane changes, braking or collision avoidance. It requires analysis of speed, distance, obstacles, and road conditions etc. Therefore, data cannot be transferred to cloud and requirement of computation is to be localized near road itself. Similarly, in smart grids, for local load balancing, fault detection and distribution management fog nodes will be required in the vicinity. Work space security can be enhanced in IIoT by utilizing fog computing [1, 2].

5.2 Characteristics of Fog Computing

The special features or characteristics of Fog computing are discussed here. Figure 5.1 depicts these characteristics of Fog computing.

T. K. Sharma et al., *Understanding Internet of Things: Evolution, Enabling Technologies and Challenges*, Studies in Computational Intelligence 1282,
https://doi.org/10.1007/978-3-032-21588-8_5

- The most important characteristics of Fog computing is it operations near an edge location. These are placed near the data origin i.e. sensors or machines which collects the data. It makes them aware of location and allows them to achieve low latency in data-processing. Hence, it provides fast, context-specific services in efficient manner [3].
- The distributed deployment of the Fog computing nodes gives it extensive geographical distribution. It is notably appealing and completely in contrast with centralized cloud architecture. It gives them capability to deliver localized and high-performance services over large area with low latency [3].
- Another important aspect of Fog computing is the capability to maintain stable communication with mobile devices. Locator and ID Separator (LISP) protocol which decouples the ID of device from its location helps in maintain communication when device transit between two different network areas [4].
- Another important aspect of Fog computing is to facilitate real-time interactions with low latency. In fog computing, data processing is carried out in proximity to IoT devices, it enables rapid service delivery, continuous operation with minimal interruption.
- Fog nodes can connect and communicate with various heterogeneous devices. Thus, it facilitates operations in diverse environments.
- IoT is amalgamation of heterogeneous devices from different vendors. It creates a situation where different data formats, protocols and standards interact with each other. Fog computing employs flexible software and networking methods that ensures compatibility amid such heterogeneity. Thus, challenges like limited bandwidth and storage capabilities, varying computational power etc. are overcome by this Fog computing's capability to ensure interoperability [6].

5.2.1 Edge Computing

Edge computing places the processing assets and information storage in the near vicinity of data source. It improves reaction time and minimizes the requirement for substantial network capacity [7]. In contrast to Fog computing, which encompasses a broader spectrum of use cases, Edge computing is particularly focused on the immediate region of the data source.

- Definition: "*Edge computing is the paradigm which enhances speed, reliability, security, and scalability by carrying out computation either on the IoT device or on the object to which it is instantly attached such as sensors, mobile phones, or IoT devices.*"
- Figure 5.2 presents the flow of information in an IoT architecture that utilizes Edge, Fog and Cloud. As mentioned in the previous paragraph, Edge computing involves the use of edge devices, including IoT devices and edge servers, to perform information processing duties on their own, hence reducing the amount of information sent to the central cloud [8].

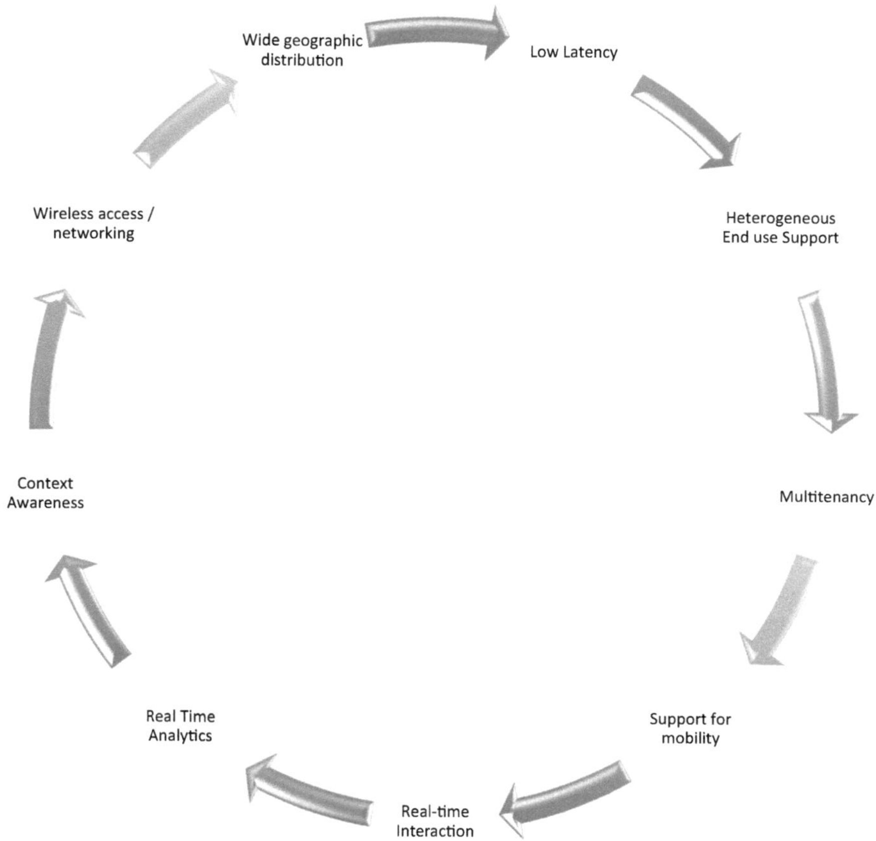

Fig. 5.1 Characteristics of Fog computing [4–6]

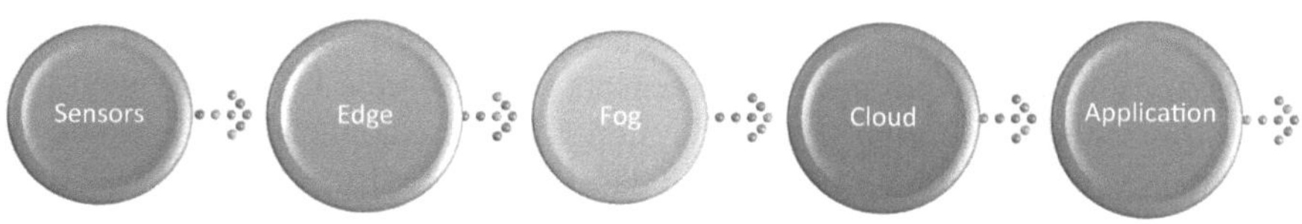

Fig. 5.2 Information flow in an architecture utilizing Edge, Fog and Cloud [7–9]

– Characteristics of Edge Computing:

- Location: processing is done on or close to IoT device.
- Processing power is very limited in edge devices.
- Latency: It exhibits minimum latency.
- Scalability: It cannot achieve scalability as each device require computation.
- Minimum bandwidth requirement.

It is ideal for scenarios that require rapid responses and where connection to the central cloud is limited or unreliable, such as video surveillance, augmented reality, and remote health monitoring [9].

5.3 Significance of Fog and Edge Computing in IoT

Both Fog and Edge computing are important for IoT ecosystem as they provide many services:

5.3.1 Low Latency and Real-Time Processing

- Immediate Data Processing: Fog and Edge computing becomes necessary when the applications are time-critical. As both processes data near its source, delay generated in communicating the data to and from cloud is eliminated. Though edge is faster as it performs computing on or immediately near the device while Fog does it slightly away.
- It also facilitates real-time analytics and decision-making for the applications like autonomous cars, industrial automation, and real-time health monitoring [10].

5.3.2 Maximizing Bandwidth Efficiency

- Minimized Data Transfer: Local analysis and compression of data with Fog and Edge computing not only reduces latency but also saves precious bandwidth required in its communication. Only essential information is transferred to cloud. It minimizes operational costs, reduces network congestion. This will help in a smart city scenario with millions of sensors or large-scale IoT deployment like environmental monitoring or traffic systems. In these applications, it would be impracticable to continuously transmit data to cloud [11].

5.3.3 Improved Security and Privacy

- Local Data Handling: Local processing of data has inherent advantage of security and privacy. As data is not transmitted, it avoids risk of exposure of information. Thereby, Fog and Edge reduce attack surface. Financial services, healthcare, and government can be the main beneficiaries [12].

5.3.4 Scalability and Reliability

- Decentralized Architecture: In Fog and Edge, workload is divided among numerous edge and fog nodes. This decentralized architecture is completely in contrast with centralized cloud systems. It avoids single point of failure and improves reliability. System uptime and resilience are also increased [11]. Fog also provides scalability which is harder to achieve with edge architecture. Distributed sensor networks and large-scale IoT installations in industrial contexts can achieve high reliability and improved uptime with Fog architecture.

5.4 Utilizations of Fog and Edge Computing in IoT

Both computing paradigms can be used to improve the speed, efficiency, and intelligence of IoT systems in multiple use-cases. It will help in reducing latency, conserving bandwidth, and enabling real-time decision-making.

5.4.1 Intelligent Urban Areas

- Traffic Management: Fog and edge nodes can be positioned at critical points such as crossings, red-lights etc. It will help in getting real-time analysis and decision making. Incidents like, immediate detection of congestion, accidents, or unusual traffic patterns can be detected in real- dynamic signal control to improve traffic flow which result reduced travel time and lower fuel consumption of traffic data obtained from cameras and sensors to enhance traffic flow efficiency and alleviate congestion [13].
- Public Safety: Large volume of data is generated by the security cameras, environmental sensors, and emergency detection devices. In Fog and edge computing, near-source processing leads to quick identification of threats, generation of alerts for quick response and improved surveillance [14].

5.4.2 Industrial Internet of Things (IIoT)

- Predictive Maintenance: Real-time data analysis of industrial equipment can be utilized in predicting machine breakdowns and equipment failures. Sensors are installed on these machines to collect temperature, pressure, vibrations and other performance parameters. Fog and edge devices can utilize this data for detecting early signs for breakdown, raise alerts and schedule maintenance. It will minimize the downtime and operational losses [15].

- By avoiding breakdowns and reducing downtime results in continuous production. Not only production is increased but production quality is maintained also. Energy consumption is optimized and minimize wastage.

5.4.3 Healthcare

- Remote Patient Monitoring: Different medical sensors and wearables detect health metrics such as blood pressure, oxygen levels, and sugar levels etc. Fog and edge nodes deployed in vicinity analyses this data to identify critical conditions, raise alert for staff to provide necessary medical intervention. This helps in providing elderly care, chronic disease management, and home-based healthcare [16].
- Telemedicine: In telehealth services, Fog and edge computing can facilitate better virtual consultations by minimizing transmission delays and providing real-time diagnostics [16].

5.4.4 Agriculture

- Precision Farming: Soil sensors, weather stations, and drones collect soil moisture, temperature, humidity, nutrient levels and environmental data. Edge or Fog systems can process this information locally to plan optimum irrigation, fertilizer scheduling and crop health issues. It will optimize operations to maximize crop yields, reduce operation cost and minimize wastage of resources [17].
- Animals Monitoring: Fog and Edge devices can help in raising alerts by analysing cattle movements, behaviour and their health indicators. It results in preventing theft and diseases. Farm management is improved and productivity is increased [18].

5.4.5 Retail

- Smart Shelves: Advanced IoT shelves equipped with weight sensors, cameras and RFID detectors can help in tracking the products. Edge devices can detect stop theft, identify shortage of items and alerting for restocking of product. In this way, automation of inventory management can be carried out with the help of edge and fog computing. It brings doen the operational costs and enhances customer satisfaction by reducing the delays [19].
- Tailored Shopping Experiences: Customers habits, preferences and movements can be analysed by collecting data from cameras and smart sensors. Analysis of this data can help in personalized promotion of products and enhancing the customer movements with better positioning of products in supermarkets [19].

5.5 Distinguishing Between Fog and Edge Computing

Both Fog and Edge computing tries to bring computation capabilities near to origin of data-source but there are still differences between both in terms of their scope, design, and applications.

5.5.1 *Scope and Architecture*

- Fog Computing: The scope for Fog paradigm exists from the edge to the cloud. It is a hierarchical structure in which nodes with computational capabilities are deployed near the edge for assisting the latency critical applications.
- Example: A smart grid system employs a hierarchical data processing approach, starting from individual meters at the edge, then moving to local substations at the fog nodes, and eventually reaching central servers in the cloud.
- Edge Computing: emphasizes the processing of data at or in close proximity to the data source, such as sensors or Internet of Things (IoT) devices. It usually includes a smaller number of intermediary nodes.
- Example: An intelligent camera system that analyses video data on-site to identify irregularities without transmitting unprocessed film to the cloud (Fig. 5.3).

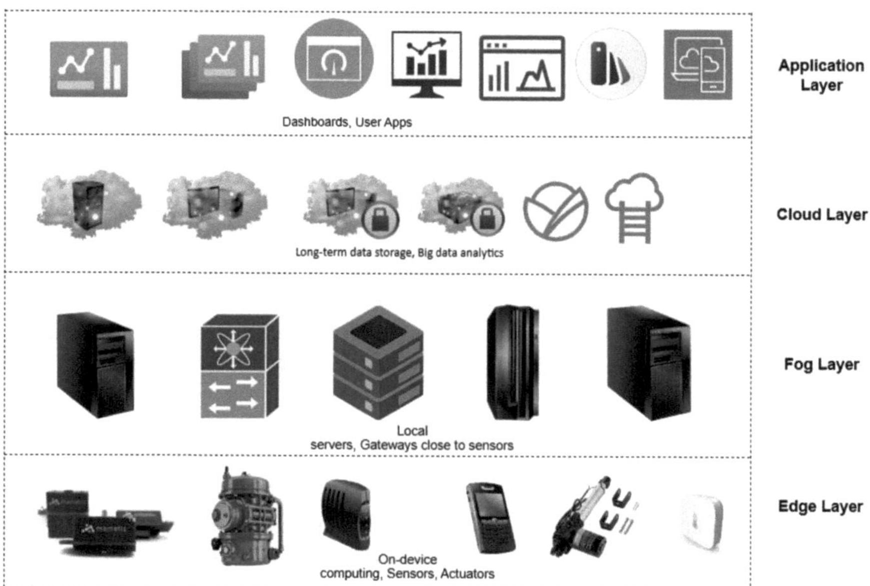

Fig. 5.3 Architectural view for cloud, fog and edge layers [20, 21]

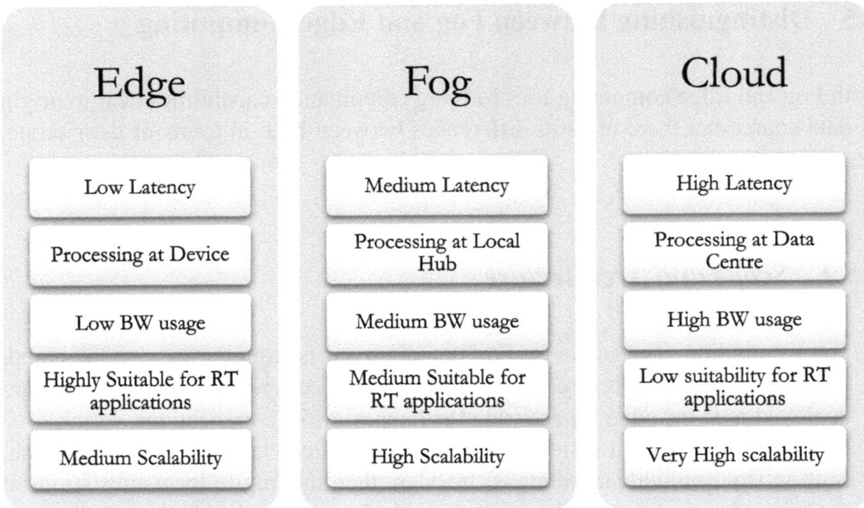

Fig. 5.4 Comparison among edge, fog and cloud [4–9, 20, 21]

Figure 5.4 presents a brief comparison among Fog, Edge and Cloud computing. This comparison will help in the right selection of architecture and capability of system for a particular use case. Let us discuss the use cases. In case of Edge, the processing will take place right at or near the IoT device. For example, in a smart city, each traffic light is equipped with cameras and sensors that detect vehicles and pedestrians. Using edge computing, the traffic light controller processes data locally to make real-time decisions like changing lights based on congestion—without needing to send data to the cloud. Its main benefits are achieving ultra-low latency, real-time response, works even with poor internet.

Second example, **Industrial Automation in a Factory,** is of **Fog Computing (local gateway or micro-data center). Here the processing takes place at** a local network node between the edge and the cloud. A manufacturing plant has multiple IoT sensors monitoring equipment. Instead of each device sending raw data to the cloud, it goes to a **fog node (like an industrial server)** on-site that aggregates, filters, and pre-processes the data. Only critical or summarized info is sent to the cloud.

It provides the b**enefits of** reduced bandwidth, allows local analytics, supports semi-real-time control.

For **Cloud Computing (remote centralized data centers), the example of Smart Home Analytics will serve as a good. The processing in this case happens at** remote cloud servers (e.g., AWS, Azure, Google Cloud). Smart thermostats, lights, and cameras send usage data to the **cloud**, where advanced machine learning models analyze the behavior patterns and optimize energy use across millions of homes. Cloud provides multiple advantages such as higher computational power, scalable storage capacity, and centralized AI-ML assistance etc.

5.5.2 *Application Scenarios and Performance*

- Fog Computing: This architecture is tailor-made for the geographically distributed IoT scenarios where low-latency but higher computational power in comparison to edge computing is required. IoT scenarios like, intelligent traffic systems across a city; Smart grids; industrial IoT with multiple machines are few such mentions where regional-real-time analytics with low latency are required. In these cases, use of Fog computing will result in reduced bandwidth consumption with advanced data aggregation and analysis.
- Edge Computing: is best suited where immediate processing of data is required, for example, autonomous vehicles needing split-second decisions, smart cameras detecting motion, wearable medical devices. These scenarios require ultra-low-latency, instant response and device-level decision making, normally in milliseconds.

5.5.3 *Illustrations of Technologies*

- Fog Computing: Some of the notable companies involved in fog computing are Cisco, Nebbiolo, and IBM. Cisco offers Cisco Fog Computing and Cisco IOx while IBM's Watson IoT Platform are popular fog services [22, 23].
- On the other hand, AWS provides IoT Greengrass, Microsoft Azure offers IoT Edge, and Google Cloud offers IoT Edge as edge computing service for IoT applications [24, 25].

5.6 Obstacles and Future Outlook

Despite multiple advantages there are multiple obstacles are also there in front of these two technologies. In order to fully harness their potential, one must address these obstacles with ingenuine solutions.

- *Security and Privacy*: Open deployment of fog and edge nodes make them potentially vulnerable to multiple threats. It becomes necessary to have strong encryption, authentication, and access control mechanism to safeguard the sensitive information. Apart from these, physical security of devices is also necessary.
- *Interoperability*: Heterogeneous devices, platforms and protocols create interoperability challenges. It becomes a necessity to achieve interoperability to attain smooth interactions and data exchange. As the IoT networks expands, more and more devices are added. Therefore, scalability is required to handle increased workload and ensure high end performance.

– *Management and Orchestration*: Fog and edge systems are distributed over large regions. Thus, challenges regarding proper load balancing, fault tolerance, updates etc. arise. Hence, effective management and coordination of these resources must be carried out for enhancing performance and reliability.

5.6.1 Outlook for the Future

– 5G/6G Connectivity: The implementation of 5G networks has provided for ultra-fast, low-latency, and highly reliable connections. Further, 6G will be launched with multiple time improvement in the existing capabilities. These updates provide for real-time applications such as autonomous cars etc.
– AI and Machine Learning: Smarter local-decision making can be achieved at Fog and Edge nodes by integrating AI-ML capabilities into these nodes. This will allow smarter decisions without involving cloud.
– Standardization: To achieve multiple goals like interoperability, security, and deployment efficiency. It is necessary that common standards and communication protocols must be developed for Fog and edge computing.
– Hybrid Architectures: It becomes must to create new hybrid architectures by combining Fog, Edge, and Cloud computing for maximizing performance, reliability, and flexibility across IoT ecosystems.

5.7 Summary

Fog and Edge computing are two technologies that have enormous impact on the capabilities and efficiency of IoT systems. These technologies position the computing and data storage near data origin for immediate processing, low-latency, minimizing and bandwidth consumption. Despite having similar objectives, they vary in multiple aspects such as location of computing resources, architecture, and applications. Both presents numerous advantages for IoT applications but also face obstacle related to scalability, interoperability and security. It is necessary to find solutions to these difficulties to fully harness the capabilities of both Fog and Edge computing. It will lead to improved efficiency, effectiveness in operations with creation of wide range of industries and uses.

References

1. Z. Mahmood, M. Ramachandran, Fog computing: concepts, principles and related paradigms, in *Fog Computing*, ed. by Z. Mahmood (Springer, Cham, 2018). https://doi.org/10.1007/978-3-319-94890-4_1

2. S. Madakam, P. Bhagat, Fog computing in the IoT environment: principles, features, and models, in *Fog Computing*, ed. Z. Mahmood (Springer, Cham, 2018). https://doi.org/10.1007/978-3-319-94890-4_2
3. K. Nagashri, S. Rajarajeswari, I.M. Imran, N.D. Shetty, Fog computing—characteristics, challenges and job scheduling survey, in *Expert Clouds and Applications*, ed. by I. Jeena Jacob, F.M. Gonzalez-Longatt, S. Kolandapalayam Shanmugam, I. Izonin. Lecture Notes in Networks and Systems, vol. 209 (Springer, Singapore, 2022). https://doi.org/10.1007/978-981-16-2126-0_52
4. A. Rayes, S. Salam, Fog computing, in *Internet of Things from Hype to Reality* (Springer, Cham, 2022). https://doi.org/10.1007/978-3-030-90158-5_6
5. R. Das, M.M. Inuwa, A review on fog computing: Issues, characteristics, challenges, and potential applications. Telematics Inf. Rep. **10**, 100049 (2023), ISSN 2772-5030. https://doi.org/10.1016/j.teler.2023.100049.
6. I. Ahammad, Fog computing complete review: concepts, trends, architectures, technologies, simulators, security issues, applications, and open research fields. SN Comput. Sci. **4**, 765 (2023). https://doi.org/10.1007/s42979-023-02235-9
7. K.S. Mohamed, Cloud, fog, and edge computing, in *Neuromorphic Computing and Beyond* (Springer, Cham, 2020). https://doi.org/10.1007/978-3-030-37224-8_8
8. D. Sabella, Principles of edge computing, fog and cloud computing, in *Multi-access Edge Computing: Software Development at the Network Edge. Textbooks in Telecommunication Engineering* (Springer, Cham, 2021). https://doi.org/10.1007/978-3-030-79618-1_1
9. A. Adel, Utilizing technologies of fog computing in educational IoT systems: privacy, security, and agility perspective. J. Big Data **7**, 99 (2020). https://doi.org/10.1186/s40537-020-00372-z
10. J. Pérez, J. Díaz, J. Berrocal et al., Edge computing. Computing **104**, 2711–2747 (2022). https://doi.org/10.1007/s00607-022-01104-2
11. S.P. Singh, A. Nayyar, R. Kumar et al., Fog computing: from architecture to edge computing and big data processing. J. Supercomput. **75**, 2070–2105 (2019). https://doi.org/10.1007/s11227-018-2701-2
12. M. Burhan et al., A comprehensive survey on the cooperation of fog computing paradigm-based IoT applications: layered architecture, real-time security issues, and solutions. IEEE Access **11**, 73303–73329 (2023). https://doi.org/10.1109/ACCESS.2023.3294479
13. V. Choudhary, Y. Singh, P. Anand, Smart traffic monitoring with fog and cloud computing, in *Emerging Technologies for Computing, Communication and Smart Cities*, ed. by P.K. Singh, M.H. Kolekar, S. Tanwar, S.T. Wierzchoń, R.K. Bhatnagar. Lecture Notes in Electrical Engineering, vol 875 (Springer, Singapore, 2022). https://doi.org/10.1007/978-981-19-0284-0_24
14. S. Khan, S. Parkinson, Y. Qin, Fog computing security: a review of current applications and security solutions. J Cloud Comp **6**, 19 (2017). https://doi.org/10.1186/s13677-017-0090-3
15. G.S.S. Chalapathi, V. Chamola, A. Vaish, R. Buyya, Industrial Internet of Things (IIoT) applications of edge and fog computing: a review and future directions, in *Fog/Edge Computing For Security, Privacy, and Applications. Advances in Information Security*, vol 83 (Springer, Cham, 2021). https://doi.org/10.1007/978-3-030-57328-7_12
16. A.K. Sarangi, A.G. Mohapatra, T.C. Mishra, B. Keswani, Healthcare 4.0: a voyage of fog computing with IOT, cloud computing, Big Data, and machine learning, in *Fog Computing for Healthcare 4.0 Environments. Signals and Communication Technology*, ed, by S. Tanwar (Springer, Cham, 2021). https://doi.org/10.1007/978-3-030-46197-3_8
17. J.A. Malik et al., Implementing fog computing in precision agriculture for real-time soil health monitoring and data management, in *Computational Intelligence in Internet of Agricultural Things. Studies in Computational Intelligence*, ed. by M.G. Sumithra, M. Sathyamoorthy, M. Manikandan, R.K. Dhanaraj, M. Ouaissa, vol. 1170 (Springer, Cham, 2024). https://doi.org/10.1007/978-3-031-67450-1_14
18. P.B. Bök, D. Micucci, The future of human and animal digital health platforms. J Reliable Intell Environ **10**, 245–256 (2024). https://doi.org/10.1007/s40860-024-00232-0

19. J. Hornik, C. Ofir, M. Rachamim, Out of the fog: fog computing-enabled AI to support smart marketing management. Manag Rev Q (2024). https://doi.org/10.1007/s11301-024-00441-0
20. B. Omoniwa, R. Hussain, M.A. Javed, S.H. Bouk, S.A. Malik, Fog/edge computing-based IoT (FECIoT): architecture, applications, and research issues. IEEE Internet Things J. **6**(3), 4118–4149 (2019). https://doi.org/10.1109/JIOT.2018.2875544
21. P.J. Escamilla-Ambrosio, A. Rodríguez-Mota, E. Aguirre-Anaya, R. Acosta-Bermejo, M. Salinas-Rosales, Distributing computing in the Internet of Things: cloud, fog and edge computing overview, in *NEO 2016. Studies in Computational Intelligence*, ed. by Y. Maldonado, L. Trujillo, O. Schütze, A. Riccardi, M. Vasile, vol. 731 (Springer, Cham, 2018). https://doi.org/10.1007/978-3-319-64063-1_4
22. D. Locke, Bringing the power of IBM Watson IoT to the Edge with Cisco (IBM), Retrieved from https://www.cisco.com/c/dam/m/ro_ro/events/2016/ciscoconnect/img/presentations/IBM.pdf. Accessed on 07 June 2025
23. F. Bonomi, *Winning with Fog Computing in the Industrial Internet of Things How to Accelerate the Journey to Productivity and Growth* (Genova, 2017), https://www.aizoongroup.com/aizooncontent/ita/Images1/5-12-2017%20Fog%20Computing%20%20GENOVA%20Nebbiolo.pdf
24. M. Hassan, The IoT cloud: Microsoft Azure vs. AWS vs. Google Cloud. Retrieved from: https://iot-analytics.com/iot-cloud/. Accessed on 07 June 2025
25. Retrieved from: https://aws.amazon.com/greengrass/. Accessed on 07 June 2025

Chapter 6
Machine-to-Machine Communication

The structure of sensors has evolved over the time. Now, they possess the capability to transmit the information and can be identified uniquely over the network. It has paved way for the realisation of IoT. A new area of research has emerged with these developments which is named as 'Machine-to-Machine (M2M)' communication. It serves as the catalyst of fourth industrial revolution. Intelligent decision lies at the heart of optimized industrial production. This capability to communicate among machines, without a human in loop, can help in faster and reliable decisions. Monitoring of machines health and conveying this information so that break downs can be projected to avoid loss of working hours and optimize repair time. Automated systems are created by the numerous such M2M exchanges. These devices with limited processing, memory and energy capabilities are quite vulnerable to many attacks. This chapter explores the basics of M2M communication. It also devolves on the uses, advantages, difficulties, and future prospects of this technology within the IoT ecosystem [1, 2].

6.1 Comprehending Machine-to-Machine Communication

The autonomous communication between sensors or small devices in a network with the help of a specialised software comes under M2M communication.

T. K. Sharma et al., *Understanding Internet of Things: Evolution, Enabling Technologies and Challenges*, Studies in Computational Intelligence 1282,
https://doi.org/10.1007/978-3-032-21588-8_6

6.1.1 *Essential Elements of Machine-to-Machine Communication*

- Sensors: These are the small devices that can detect and measure different physical quantities such as temperature, pressure, or motion etc. and transmit them over the network.
- Actuators: The devices that are used to convert a command into action. It is achieved by converting energy into mechanical work e.g. stepper motors or DC torque motors.
- Networks: Any type of network, such as cellular, Wi-Fi, and Bluetooth, that can enable such communication among devices.
- Data Processing Units: Those devices or systems which help in the processing, calculating and manipulating the data collected by the sensors.
- Software: The programs or applications that help in managing and controlling M2M communication procedures e.g. oneM2M-compliant CCSP and CoSMiC.

6.1.2 *Types of M2M*

There isn't a single "official" classification of M2M types, but they generally fall into five major dimensions:

1. Connection type (wired vs wireless)
2. Network type (cellular, capillary, LPWAN, satellite)
3. Application area (industrial, automotive, healthcare, etc.)
4. Device role (sensor, actuator, gateway)
5. Power/data class (high vs low)

Based on Communication Medium

The two classifications based on communication medium used are provided in table below (Table 6.1):

Based on Network Type

Based on network type, M2M can be divided in four different classifications (Table 6.2):

Table 6.1 M2M classification based on communication medium

Type	Description	Examples
Wired M2M	Devices communicate via physical cables	Ethernet-connected industrial systems
Wireless M2M	Uses wireless signals for communication	IoT sensors using Wi-Fi, cellular, etc

Table 6.2 M2M classification based on network type [1–5]

M2M type	Explanation
Capillary M2M	Short-range connection via a gateway (e.g., Zigbee, Bluetooth, Wi-Fi)
Cellular M2M	Direct connection using cellular networks (2G, 3G, LTE, NB-IoT, 5G)
Satellite M2M	Remote connectivity via satellite (used in maritime, mining, rural areas)
LPWAN M2M	Low-power wide-area networks like LoRa, Sigfox, NB-IoT for long-range low-data use

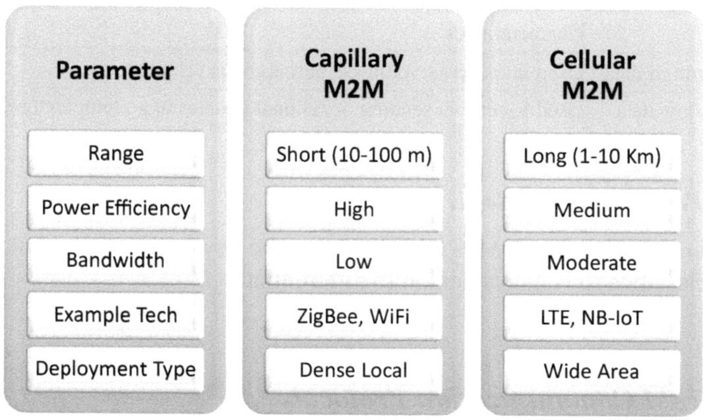

Fig. 6.1 A comparison between capillary and cellular M2M [4, 5]

Figure 6.1 provides the key differences between capillary and cellular M2M.

Based on Use Case/Application

Use case also categorize the M2M into five different categories (Table 6.3):

Based on Device Role

Further four different categories can be created based on the role of devices (Table 6.4):

Table 6.3 M2M classification based on use cases

Application type	Used in
Industrial M2M	Manufacturing, robotics, factory automation
Automotive M2M	Vehicle tracking, telematics, diagnostics
Healthcare M2M	Remote patient monitoring, smart wearables
Smart City M2M	Traffic lights, environmental sensors, public safety systems
Consumer M2M	Smart home devices, fitness trackers, connected appliances

Table 6.4 M2M classification based on device role

Type	Function
Sensor M2M	Collects data from the environment (e.g., temp, GPS)
Actuator M2M	Takes action based on a signal (e.g., open valve, turn fan)
Gateway M2M	Aggregates data from sensors, may relay to cloud
End-to-End M2M	Devices that directly communicate with cloud/app

Table 6.5 M2M classification based on power and data needs

Type	Characteristics
High power/high data	Used in video surveillance, autonomous vehicles
Low power/low data	Used in sensors sending occasional updates (e.g., temperature, location)

Based on Power and Data Needs

The last classification of M2M is done based on the power and data (Table 6.5): These classifications help a person to differentiate.

6.1.3 M2M Communication Protocols

- MQTT (Message Queuing Telemetry Transport) is a messaging protocol specifically developed for devices with limited resources and networks with low bandwidth and high latency [6].
- CoAP (Constrained Application Protocol): it is an important protocol that works on client/server model based on RESTful architecture. It identifies the network objects through a unique resource identifier (URI). As this protocol is tailor made for low-power devices; it employs asynchronous messages without acknowledgements [6]. With a 4-byte header and variable 0–8 bytes payload and token size, it offers a lightweight, low overhead asynchronous communication.
- HTTP/HTTPS: This protocol can be a secure method for M2M applications that requires to communicate with remote servers. These are applications are those which have fewer resource limitations [6, 7]. It is fairly easy for device developers to create an HTTPS-based Internet of Things protocol that can connect to any kind of web server or backend application by using an equipment that has a minimal HTTP client library and a short SSL stack. An application server is usually needed for the backend infrastructure since, while a web server can process HTTPS requests, it usually is unable to process the business logic needed to manage the data transferred to and from the connected devices.
- AMQP (Advanced Message Queuing Protocol) is a protocol for message-oriented middleware, ensuring dependable and secure communication, particularly for

complex distributed and mission-critical systems. Interoperability and dependability are its main advantages. AMQP is appropriate for large-scale applications such as the Ocean Observatories Initiative and India's Aadhar project because it uses queues to transmit transactional messages between servers, processing thousands of dependable queued transactions.

6.2 Significance of Machine-to-Machine Communication in the Internet of Things

M2M communication is critical for the function and success of Internet of Things (IoT) systems. Automation of tasks is enabled through fast data transfer with M2M communication. Efficiency of operations is increased for many applications in this way.

6.2.1 Enabling Automation

M2M communication has made autonomous operations of devices for the execution of tasks based on real-time data without human interference. Amany IoT applications such as smart homes, healthcare monitoring, and industrial automation utilize M2M communication for different tasks. In smart home systems, sensors detect motion and communicates this information to hub. The hub automatically switches lights on and off based upon the information from these sensors.

6.2.2 Improving Efficiency

Normally, a technician in reads a sensor and manually enter that data so that different decisions can be made such as '*switching ON the pump*' or '*minimizing the power*'. M2M technology enables devices to communicate automatically. Hence, analysis is done in real-time to take appropriate action. It limits human intervention to high-level supervision or exceptions. By improving efficiency, this raises the efficacy of systems and processes. For example, in smart grids, smart meters are utilized to measure power consumption. These devices identify the power patterns of industries and households and transfer this information to smart grid controller using M2M communication. These patterns are then used by smart transformers and substations to re-route the energy to high demand region from low-consumption area. Thereby, optimizing the distribution of power in smart grids. Similarly, Industrial IoT uses M2M communication for predictive maintenance. The readings from vibration and temperature sensors are communicated to controller. It shuts down the machine before breakdown and minimizes the shut down time.

6.2.3 *Real-Time Monitoring and Control*

In dynamic environment, IoT systems require continuous monitoring to make the real-time adjustments as per conditions. If a patient is in ICU, his vitals are continuously being monitored with the help of different sensors. Suppose the oxygen-level drops below normal, this is informed to life-support system via M2M communication. As a response, the oxygen supply is increased without the pressing the button by a nurse. The situation can also be updated to the patient's dashboard as well.

6.2.4 *Utilizing Data to Make Informed Decisions*

M2M communication allows the collection and communication of data. This data can be used to find various hidden patterns and trends. A detailed analyses can generate useful insights that can be very helpful for making well-informed decision. For example, in agriculture IoT systems, humidity, soil moisture, temperature etc. are collected at a central server using M2M communication. Upon analyses, this data provides information about the dry areas, amount of water and fertilizers required, pattern of irrigation that needs to be followed. In this way, real-time decisions can be made for irrigating the dry areas, scheduling the spray of fertilizers and irrigation. This can help farmers to optimize water usage, planning harvest for markets and identify pest-risks.

6.3 Utilizations of Machine-to-Machine Communication in the Internet of Things

A number of Internet of Things (IoT) applications are improved with help of M2M communication by improving the automation, effectiveness, and intelligence in multiple domains such as cities, homes, agriculture and industries.

6.3.1 *Intelligent Urban Areas*

- Traffic Management: Sensors and cameras work together to optimize the movement of vehicles, decrease traffic congestion, and improve public safety [8]. This real-time information can be useful in offering better alternative for vehicles.
- Public Utilities: Smart meters and sensors are used to monitor the consumption of utilities, allowing for effective management of resources and minimizing wastage [9]. Based on the information and notifications they provide, smart electricity meters assist energy users in understanding how to save energy. One active and

continuing research topic to produce energy saving solutions is advanced metering infrastructure (AMI).

6.3.2 *Healthcare*

- Remote Patient Monitoring: The intelligent sensors and devices can be positioned on human body to capture the vitals of a person. These parameters are transmitted using M2M communication to healthcare providers for ongoing monitoring [10]. As the body area networks are created by very small sensors, M2M communication becomes important because of its energy efficient nature.
- Medical Device Integration: M2M communication facilitates the integration of different medical equipment, enabling smooth data sharing and coordinated healthcare [11]. Near-zero error tolerance is essential for sophisticated and costly medical equipment that treat life-threatening illnesses. It is advised to use M2M technologies for proactive maintenance. Usage data from connected medical devices can be streamed to a secure cloud for ongoing, real-time analysis and monitoring. This makes it possible to anticipate and stop future malfunctions, which makes it easier to replace worn parts on time and drastically lowers maintenance expenses for device owners.

6.3.3 *Agriculture*

- Precision Farming: It is type of farming that employs IoT devices, M2M communication, sensors etc. for real-time monitoring of farm conditions with an aim to increase yield and minimize wastage of resources. This is achieved by well-informed decisions based on data collected from farm. Apart from temperature, humidity and moisture, soil nutrient sensors, drone-cameras and weather stations can also be utilized to create a real-time, self-regulated system. This system can help in managing irrigation and fertilizers schedules, pest-risk identification, type and quantity of fertilizers to be sprayed, crop-yield estimation etc. All these can help the farmers to make strategies for accurate and targeted farming techniques [11, 12].
- Livestock Management: For illness prediction, breeding management and other important tasks for livestock IoT systems can be developed. These systems employ M2M communication with the help of appropriate sensors to monitor animal's health, location, behaviour and movements. Different sensors, like wearable collars, tags or ingestible sensors are used to collect this data. If the body temperature of a cow increases and its movements slow down then collar can send an alert to owner's phone. The cattle's grazing can be managed using the GPS tracker collar which can act as virtual fence for these cattle [11, 12].

6.3.4 Intelligent Residences

In intelligent residences, M2M communication between different sensors like temperature, motion, light etc. with devices and systems is used for automation and energy-efficient operations to make the living environment comfortable, safe and better for its inhabitant.

- Home Automation: In this scenario, motion sensors can identify if someone is entering in the room. This will trigger smart lightning system, temperature sensor and camera. Smart lightning system will turn on the light. Temperature sensor will signal thermostat to help adjust the air conditioner (AC). While cameras can start recording. In case of suspicious activity, an alert can be sent to phone and smart lock can be activated to secure the doors. On detecting no activity, lights and air conditioning will be turned off, thereby, saving electricity. In this way, automation of different aspects of home will make the living experience much better and safe. All this will be achieved using M2M without human intervention by enabling automated control and monitoring [13].
- Energy Management: Apart from saving energy, when the home is not occupied by the people. Smart meters can identify the peak demand for energy in house. It can command other intelligent appliances to schedule their tasks when the tariff rates are low or when the demand/load is minimum. Using the light sensors, smart windows and lightning systems can be managed to minimize the energy consumption by controlling lights and AC [9].

6.3.5 Industrial Internet of Things (IIoT)

- Predictive Maintenance: By utilizing sensors installed on machinery, the performance can be continuously monitored and any abnormalities may be detected. This diagnostic information enables the implementation of predictive maintenance strategies, which in turn minimizes the amount of time that the machinery is out of operation. Real-time M2M data from connected medical devices enables the analysis of usage statistics to identify patterns. Integrating this data with a medical device manufacturer's production department provides insights into part wear, inventory needs, and design improvements for efficiency and reduced errors. This advance information, facilitated by M2M data integration with ERP systems, streamlines production, lowers inventory costs, and supports a lean and agile supply chain [6, 14].
- Process Automation: M2M communication enables the automation of industrial processes, leading to enhanced productivity and decreased occurrence of human mistake. [6]

6.4 Advantages of Machine-to-Machine Communication in the Internet of Things

Apart from encouraging innovation and improving productivity, one can list many other benefits of the use of M2M communication in the IoT based on the previous discussion in this chapter.

6.4.1 *Reduction in Expenses*

- Operational Efficiency: Labour costs are reduced through automation as lesser manual interference is needed. It results in cost savings.
- Predictive Maintenance: The role of M2M communication in the early prediction of problems helps in reducing possible down-time as well as repair-time. It cuts down the expenditure on maintenance. Low down-time and quick turn-around leads to improved productivity.

6.4.2 *Enhanced Quality of Service*

- Real-Time Data: As M2M enables real-time feedback from the environment and system. This data leads to quick and real-time response to any problem that arises. Hence, quality of service improves for the user.
- Customization and Personalization: The user inputs create the opportunities for designing more specialized services and solutions that suits user's requirement. Real-time adaptations increase user happiness.

6.4.3 *Scalability*

- Versatile Deployment: The capability of M2M communication devices to utilize different technologies like Wi-Fi, 4G/5G, LoRa, Bluetooth etc. enables them to accommodate the changing requirements and requests. Self-configuring and self-updating properties of these nodes make these suitable for scalable operations as per the requirements.
- Global Connectivity: As discussed in previous point, M2M networks can utilize different technologies for communication. Using 5G or satellite communication let these networks to connect with on global scale. In this way, seamless connectivity and coordination can be ensured to support international logistics and supply chains.

6.4.4 *Improved Security*

- Secure Protocols: M2M communication is used for critical data and mostly unattended. Most of the M2M communication protocols have features to ensure confidentiality, authentication and privacy etc. to safeguard the data.
- Monitoring and Alerts: The AI algorithms for anomaly detection can be deployed to identify security breaches. This type of continuous monitoring can help raise real-time alert so that swift actions can be taken against potential threats.

6.5 Difficulties of Machine-to-Machine Communication in the Internet of Things

Just like two faces of coin, some difficulties are also associated with M2M communication in IoT. These obstacles require innovative resolution.

6.5.1 *Ensuring the Security and Privacy of Data*

- Susceptibility to Attacks: Deployment of large number of M2M communication devices increases the surface area for attacks. Low computing power of these devices provides for limited security measures. Such systems become highly vulnerable due to lack of regular updates. [16].
- Privacy Concerns: M2M networks gathers huge amounts of personal data. These data are transmitted over various networks making them more prone to attacks. As users have limited visibility and control over the data. This data can be utilized by companies without their consent. It creates concerns over the privacy of data.

6.5.2 *Interoperability*

- Standardization: Multiple vendors with their own devices and rules may lead to incompatibility. It may also make data in-consistent and reduce the reliability of its communication. The process of standardization will ensure mass deployment without customization, thereby, reducing cost and complexity.
- Interoperability: Once the standardization achieved, it will ensure same protocols, data-formats and communication procedures. It will allow for the interoperability of devices from different manufacturers. It will result in seamless collaboration and automation.

6.5.3 *Energy Efficiency*

- Small size of devices and their energy capabilities make energy efficiency is a critical factor for M2M communication. Every node in low-power communication networks has a limited amount of energy. Therefore, mechanisms need to be added to conserve energy in such systems. Addition and deletion of nodes make it challenging to save energy. Designing of efficient neighbour discovery protocols can lead to effective solution of this problem [15]. Similarly, controlling the sleep time is also important.

6.5.4 *Enhancing Network Reliability*

- Access Challenges: Due to small size of sensor nodes, the battery carrying capacity is also limited. It also results in low transmission power. Therefore, maintaining a reliable communication link becomes very difficult for such small networks. The biggest challenge is to ensure consistent and uninterrupted network access, particularly in remote or difficult-to-reach areas.
- Bandwidth Constraints: Though the sensors transmit data in Kilobits or few bytes of data only 3–4 times in an hour. The enormous number of sensors deployed in an area can results in few TBs of data in a day. Therefore, effective restrictions on the amount of data transmitted may help in avoiding congestion and delays.

6.5.5 *Scalability*

- Infrastructure Requirements: Scalability is an important issue in IoT applications. Accommodation of the newly added interconnected devices along with the surging data traffic are two main requirements for scalability. Therefore, a robust infrastructure that must fulfil these two demands must be developed.
- Allocation of resources: Designer must provide the sufficient computational and storage resources that can meet the requirements of M2M communication.

6.6 Outlook for the Future

The future of machine-to-machine (M2M) communication in the Internet of Things (IoT) is poised for significant breakthroughs, propelled by technological progress and growing acceptance across several industries.

6.6.1 Artificial Intelligence and Machine Learning

The advancement in the field of AI and ML has made it possible to analyse massive amount of data. It can detect patterns and trends and help in predicting events. In this way, better examination of M2M data result in comparatively improved decisions.

6.6.2 Edge Computing

- Localized Processing: Edge computing provides the capability to process the data near the source like nearby gateway or embedded processor. As it does not send data to cloud, both latency, bandwidth consumption decreases. The response time also reduces which suits the need for real time applications [16].
- Hybrid Architectures: As complex applications are being developed, need for hybrid architectures are increased. It converges the cloud and edge model into one. The real-time, latency-sensitive tasks such as emergency response are transferred to edge while in computationally intensive tasks such large-scale analytics are transferred to cloud. For example, in hybrid architecture traffic lights can be controlled by edge computing while city-wide long term traffic flow can be analysed and optimized for better traffic [16].

6.6.3 Blockchain Technology

- Enhancing Data Security: Blockchain is a distributed ledger technology where only trusted devices can participate to conduct verified transactions. It provides security, transparency and enhances trust through tamper-proof decentralized computation. There is not a single point of failure. For example, in smart grid scenario, the identities and other credentials regarding the smart meters can be stored in the blockchains. In case a user sells energy back to grid the transaction can be recorded utilizing blockchain technology [17]. Without validation by the network, no one can carry out such transactions. As only trusted nodes with necessary cryptographic keys can make such transactions, data security and privacy in M2M communication is improved [18].
- Transparent and Verifiable: All the transactions in a blockchain network are immutable as these are recorded in a ledger and timestamped. It allows the audit of transaction by all the participating nodes. Thereby, enhancing the trust and accountability. In healthcare IoT systems, M2M devices like patient monitoring devices, diagnostic machines, and hospital servers exchange sensitive medical devices. End-to-end visibility is ensured for all the participants. It prevents falsification of records and increases the accountability [17–19]. Access and modification information of medical data can be verified by audit. Private information

is also not exposed which also ensures compliance of regulations such as HIPAA or GDPR etc.

6.6.4 5G and 6G Connectivity

- Enhanced Data Transmission: The capability of the rapid and efficient transmission of data can be achieved by utilizing the 5G and 6G technologies. Integration of 5G has been already achieved which resulted in numerous new applications. Data rates offered by 6G are expected to be 50–100 times more as compared to 5G. This exponential improvement in data rates will make the way for the development of powerful and advanced applications of M2M communication [20].
- Large-scale IoT Deployment: Availability of transmission speeds in Tera bits per second and bandwidths in GHz range will allow for the deployment of a large-scale IoT devices. High bandwidth will also provide better device management.

6.7 Summary

M2M communication acts as a catalyst for the development of various IoT applications. By removing human interference, it boosts automation. Effectiveness of decision making is enhanced in diverse applications by the use of M2M communication. Its significance is highlighted by the huge amount of instantaneous data interchanges between devices. Though, it faces many difficulties regarding data security, interoperability, and network stability. The integration of 5G and 6G networks, blockchain technologies, A and ML etc. portrays a positive outlook for the future. Overall, combination of M2M communication with IoT points towards the many important solutions and transformations for several industries in coming future.

References

1. V. Rohokale, R. Prasad, Cyber security for intelligent world with internet of things and machine to machine communication. J. Cyber Secur. Mobility **4**(1), 23–40 (2015). River Publishers https://doi.org/10.13052/jcsm2245-1439.412
2. M. Condoluci, G. Araniti, T. Mahmoodi, M. Dohler, Enabling the IoT machine age with 5G: machine-type multicast services for innovative real-time applications. IEEE Access **4**, 5555–5569 (2016). https://doi.org/10.1109/ACCESS.2016.2573678
3. G. Yi, J.H. Park, S. Choi, Energy-efficient distributed topology control algorithm for low-power IoT communication networks. IEEE Access **4**, 9193–9203 (2016). https://doi.org/10.1109/ACCESS.2016.2630715
4. A. Bartoli, M. Dohler, A. Kountouris, D. Barthel, 12—advanced security taxonomy for machine-to-machine (M2M) communications in 5G capillary networks, in *Machine-to-Machine (M2M) Communications*, ed. by C. Antón-Haro, M. Dohler (Woodhead Publishing,

2015), pp. 207–226, ISBN 9781782421023. https://doi.org/10.1016/B978-1-78242-102-3.00012-5.
5. A. Alexiou, A. Gotsis, 10—packet scheduling strategies for machine-to-machine (M2M) communications over long-term evolution (LTE) cellular networks, in *Machine-to-Machine (M2M) Communications*, ed. by C. Antón-Haro, M. Dohler (Woodhead Publishing, 2015), pp. 173–186, ISBN 9781782421023. https://doi.org/10.1016/B978-1-78242-102-3.00010-1.
6. M. Cavalcanti, H. Costelha, C. Neves, Industry 4.0 machine-to-machine communication protocols and architectures on the shop floor, in *Progress in Digital and Physical Manufacturing. ProDPM 2021. Springer Tracts in Additive Manufacturing*, ed. by J.O. Correia Vasco, et al. (Springer, Cham, 2023). https://doi.org/10.1007/978-3-031-33890-8_19
7. T. keophilavong, Widyawan, M.N. Rizal, Data transmission in machine to machine communication protocols for internet of things application: a review, in *2019 International Conference on Information and Communications Technology (ICOIACT)* (Yogyakarta, Indonesia, 2019), pp. 899–904. https://doi.org/10.1109/ICOIACT46704.2019.8938420
8. D.E. Ruiz-Guirola, O.L.A. López, S. Montejo-Sánchez, Modeling IoT traffic patterns: Insights from a statistical analysis of an MTC dataset. Expert Syst. Appl. **272** (2025). 126726, ISSN 0957-4174, https://doi.org/10.1016/j.eswa.2025.126726
9. M. Dujak, V. Parać, M. Đurasević, A. Herić, Machine-to-machine communication as key enabler in smart metering systems, in *2013 36th International Convention on Information and Communication Technology, Electronics and Microelectronics (MIPRO)* (Opatija, Croatia, 2013), pp. 409–414
10. E. Kartsakli, A.S. Lalos, A. Antonopoulos, S. Tennina, M. Di Renzo, L. Alonso, C. Verikoukis, 20—Machine-to-machine (M2M) communications for e-health applications, in *Machine-to-Machine (M2M) Communications*, ed. by C. Antón-Haro, M. Dohler (Woodhead Publishing, 2015), pp. 375–397, ISBN 9781782421023. https://doi.org/10.1016/B978-1-78242-102-3.00020-4.
11. S.H. Shin, R. Kamal, R. Haw, S.I. Moon, C.S. Hong, M.J. Choi, Intelligent M2M network using healthcare sensors, in *2012 14th Asia-Pacific Network Operations and Management Symposium (APNOMS)* (Seoul, Korea (South), 2012), pp. 1–4. https://doi.org/10.1109/APNOMS.2012.6356067
12. R. Akhter, S.A. Sofi, Precision agriculture using IoT data analytics and machine learning. J. King Saud Univ. Comput. Inf. Sci. **34**(8), 5602–5618 (2022), ISSN 1319-1578. https://doi.org/10.1016/j.jksuci.2021.05.013
13. M. Javaid, A. Haleem, R.P. Singh, R. Suman, Enhancing smart farming through the applications of Agriculture 4.0 technologies. Int. J. Intell. Netw. **3**, 150–164 (2022), ISSN 2666-6030. https://doi.org/10.1016/j.ijin.2022.09.004.
14. S. Despoudi, K. Spanaki, O. Rodriguez-Espindola, E.D. Zamani, From industry 4.0 to agriculture 4.0, in *Agricultural Supply Chains and Industry 4.0* (Palgrave Macmillan, Cham, 2021). https://doi.org/10.1007/978-3-030-72770-3_3
15. G. Yi, J.H. Park, S. Choi, Energy-efficient distributed topology control algorithm for low-power IoT communication networks. IEEE Access. Special Section on Intelligent Sensing on Mobile and Social Media Analytics, Digital Object Identifier (2016) https://doi.org/10.1109/ACCESS.2016.2630715
16. J. Leng, Z. Chen, W. Sha, S. Ye, Q. Liu, X. Chen, Cloud-edge orchestration-based bi-level autonomous process control for mass individualization of rapid printed circuit boards prototyping services. J. Manuf. Syst. **63**, 143–161 (2022), ISSN 0278-6125. https://doi.org/10.1016/j.jmsy.2022.03.008.
17. S. Yin, J. Bao, Y. Zhang, X. Huang, M2M security technology of CPS based on blockchains. Symmetry **9**(9), 193 (2017). https://doi.org/10.3390/sym9090193
18. P. Vilchez, S. Jacques, F. Freitag, R. Meseguer, LoRaTRUST: blockchain-enabled trust and accountability service for IoT data. Electronics **12**(9), 1996 (2023). https://doi.org/10.3390/electronics12091996
19. B. Shala, U. Trick, A. Lehmann, B. Ghita, S. Shiaeles, Novel trust consensus protocol and blockchain-based trust evaluation system for M2M application services. Internet of Things **7**, 100058 (2019), ISSN 2542-6605. https://doi.org/10.1016/j.iot.2019.100058.

20. S.S. Dhanda, T.K. Sharma, B. Singh, P. Jindal, D. Panwar, 6G-enabled internet of medical things, in *Soft Computing: Theories and Applications. Lecture Notes in Networks and Systems*, vol. 627, ed. R. Kumar, A.K. Verma, T.K. Sharma, O.P. Verma, S. Sharma (Springer, Singapore, 2023). https://doi.org/10.1007/978-981-19-9858-4_7

Chapter 7
Software-Defined Networking in the Internet of Things

As per CISCO, IoT is the first evolution of internet with which network of networked has become network of things. It now encompasses various objects such as sensors, actuators, smart appliances, and industrial equipment etc. With this exponential growth of IoT devices, the management of these devices becomes a cumbersome task. Software-Defined Networking (SDN) is a potential solution for this problem which presents other possible advantages along with this. It offers a flexible, programmable, and centralized approach to network management. This chapter will examine the incorporation of SDN with IoT, exploring its structure, advantages, difficulties, and potential future developments.

7.1 Introduction to Software-Defined Networking

SDN is a new paradigm in which dynamic and flexible administration is carried out by separating of data and control planes in network. It follows a hierarchical architecture and centralised management of switches and routers. In addition to that network virtualisation techniques helps in abstraction of physical network. It also allows functional improvements and hardware device updates through an open programmable interface. These characteristics of SDN helps it in meeting the continuously changing demands and coordination of resources to meet specific needs and objectives [1].

7.1.1 Traditional Networking and Need for Software-Defined Networking

Traditional networking architectures are static by nature and have and decentralized architecture. It becomes quite difficult for them to manage the explosive growth of

T. K. Sharma et al., *Understanding Internet of Things: Evolution, Enabling Technologies and Challenges*, Studies in Computational Intelligence 1282,
https://doi.org/10.1007/978-3-032-21588-8_7

IoT devices due to dynamic nature of IoT networks where devices are continuously entering and leaving the network. Due to distributed architecture traditional networks, networking devices manage their own control logic [2].

Key Features of Traditional Networks:

- Decentralized Control—Each router/switch makes its own forwarding decisions.
- Static Configurations—Network policies are manually configured on each device.
- Limited Scalability—Adding new devices increases complexity.
- Difficult Traffic Management—No real-time optimization of network paths.

Challenges in Traditional Networking:

- Complex Network Management—Manually configuring routers/switches is time-consuming.
- Lack of Programmability—Network policies cannot adapt dynamically.
- Slow Response to Traffic Changes—No centralized mechanism for optimizing data flows (Table 7.1).

These challenges emphasize the need to develop a scalable, secure, programmable network. It should be easy to manage and exhibit low latency.

Table 7.1 Key Challenges in traditional IoT networks [1–10]

Challenge	Description	Example
Scalability issues	IoT networks must handle millions of devices but traditional networks lack centralized control	*Smart cities* with thousands of IoT sensors require *flexible network scaling*
Complex network management	IoT devices use *diverse protocols* (MQTT, CoAP, HTTP), making *management difficult*	Industrial IoT has multiple vendors with different device configurations
Latency and performance bottlenecks	*High network traffic from IoT* devices leads to *latency and congestion*	*Autonomous vehicles* need *real-time data processing,* which traditional networks struggle to deliver
Security risks and vulnerabilities	IoT devices often have weak encryption and passwords, making them prone to attacks	Mirai Botnet (2016) hijacked thousands of IoT cameras for DDoS attacks
Lack of programmability	Traditional networks use *static configurations,* making it difficult to adapt to changing IoT demands	*Smart healthcare* requires *dynamic bandwidth allocation* for critical patient monitoring

7.1.2 Software-Defined Networking and Its Architecture

Network monitoring and configuration has been simplified by SDN. Open Flow (OF) protocol, was developed by Stanford University via a southbound API it communicated with controllers and data planes [3]. OF is used to install forwarding rules on network devices. SDN also has a management plane, allowing administrators and network engineers to deploy network applications like load balancers and routing protocols. The latest version, 1.6, includes new functions [4].

SDN is a cutting-edge networking approach that divides the control plane and data plane in network devices. This segregation enables the consolidation of network administration, where a controller based on software governs the flow of traffic, while network equipment (such as switches and routers) only transmits packets according to the controller's directives. Hence, one can define SDN as follows:

> Software-Defined Networking is an innovative network architecture that separates the control plane (decision-making) from the data plane (packet forwarding), enabling centralized control and programmability.

The architecture of SDN is presented in Fig. 7.1. It details the architecture in to 3 main parts.

– **Control Layer and Control Plane (Centralized Network Management)**

The control plane is the centralized component that is responsible for network intelligence and decision-making. The system operates on SDN controllers, which oversee the network's behaviour and policies. It is the brain of SDN [3, 4].

- It resides in the SDN Controller, making global decisions for network traffic.
- It optimizes routing, security policies, and traffic flow dynamically.

For example, if a network link fails, the controller reroutes traffic automatically instead of relying on slow traditional routing protocols.

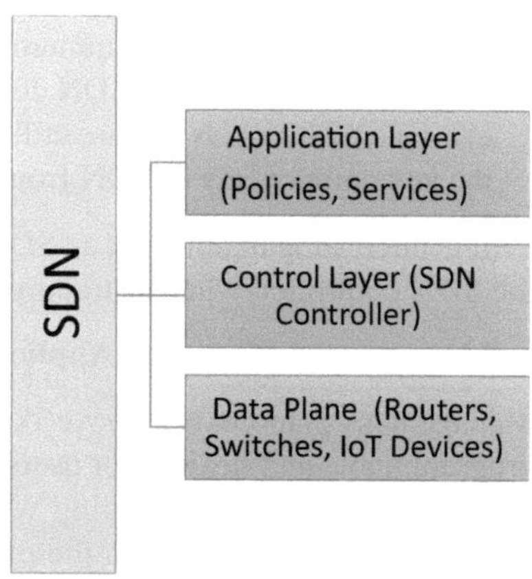

Fig. 7.1 SDN architecture

The SDN control layer is the core of the network structure, interacting with the infrastructure and application layers via southbound and northbound APIs [5]. It consists of four main modules: link discovery, topology manager, decision making, and flow manager. The link discovery module builds the network topology using two protocols: Link Layer Discovery Protocol (LLDP) and Broadcast Domain Discovery Protocol (BDDP). The topology manager manages real-time connection and working status of SDN switches. The decision-making module handles data forwarding behaviours, while the flow manager distributes flow tables, either actively or passively. In complex networks, multiple SDN controllers can be used with east–west bound interfaces [6, 7].

– **Data Plane (Forwarding Devices)**

The data plane refers to the distributed element including network devices responsible for packet forwarding according to the instructions provided by the control plane [8].

- Consists of physical and virtual switches/routers that forward packets based on SDN Controller instructions [9].
- Uses OpenFlow rules for packet forwarding [10].

For example, Open vSwitch (OVS) is a popular SDN-enabled switch that forwards packets under controller instructions.

- Southbound APIs: These are interfaces that enable communication between the control plane and data plane. OpenFlow is a commonly utilized southbound Application Programming Interface (API) [11].
- Northbound APIs: These are interfaces that facilitate the interaction between applications and services with the control plane. They enable the automation and programmability of network functions. These API also provide a hawk-eye view of the entire network [11].
- Westbound APIs and Eastbound APIs—These API are utilized to maintain the communication and coordination between distributed SDN controllers to enable the better collaborative decision making. While Westbound interfaces are responsible for the traditional networks and SDNs. SDN-IP and RouteFlow are the protocols that supports such communication. On the other hand, eastbound interfaces are responsible for inter-SDN communication. ONIX, SDNi and hyperFlow are example of protocols that are utilized for the purpose [11, 12]. One can summarize the key components of SDN from the table (Table 7.2):

It is interesting to note that an SDN Firewall can be updated dynamically through the SDN controller, whereas traditional firewalls require manual reconfiguration.

– **Management Plane and Application layer**

Management plane exists above control plane. Its main task is to make network programmable and flexible for users. Northbound APIs are utilized by users' applications built for communicating with the controller. These applications have the authority to make changes in the physical networks upon which it works with the help of a high-level language. This is facilitated by the control layer. That is why,

Table 7.2 Key components of SDN

Component	Description
Control plane	Centralized SDN controller that manages traffic flow
Data plane	Network devices (switches, routers) that forward packets based on SDN rules
Southbound API	Connects the SDN controller to network devices
Northbound API	Allows applications such as firewalls, monitoring tools etc., to interact with the SDN controller

one may consider SDN as a "platform as a service" model in context of networking [3, 12].

7.1.3 *Main Characteristics of SDN*

- Centralized Management: Its main functions include configuration, monitoring, troubleshooting, and policy enforcement which are to controlled by a single unified platform. Centralized management provides sstreamlined configuration, faster troubleshooting and complete view of the entire network through this single unified platform which is also known as dashboard. It reduces complexity, improves consistency and timing, easy control, especially in large environments.
- Programmability: It highlights the configuration of network via a software. Programmability allows for dynamic configuration, automation and quick adaptation. It results in lesser errors, fast deployment and scalable adaptation without human intervention.
- Scalability: It refers to the ease with which a network can expand and shrink by addition and removal of devices, services, and users. SDN allows for abstraction of hardware complexity, flexible resource allocation and expansion of topologies. SDN enabled scalable networks can address high-demand, avoid bottlenecks and expand as the network grows.
- Flexibility: It highlights the networks capability to adopt new services, applications and protocols. SDN enables network to be flexible without changing the underlying hardware. The changes to adopt these services, protocols etc. are made in software only. Virtualization, SDN controllers, APIs etc. allows for these changes. Flexibility of network allows for a future proof design that can boost innovation with low costs.

In SDN, an SDN Controller (e.g., ONOS, OpenDaylight) dynamically adjusts network traffic instead of requiring manual router reconfigurations [6, 8]. Table 7.3 and Fig. 7.2 provides a detailed comparison between SDN and traditional networking features and structure wise respectively. It can be observed that SDN is a more suited to the needs of IoT. On the other hand, traditional networking architecture was found lacking on different aspects of IoT.

Table 7.3 SDN versus traditional networking [2, 4–9]

Feature	Traditional networking	Software-defined networking (SDN)
Network control	Distributed (device-based)	Centralized (controller-based)
Configuration	Manual, device-by-device	Automated, software-driven
Scalability	Limited	High scalability
Security	Static security rules	Dynamic security enforcement
Adaptability	Hard to modify	Flexible and programmable

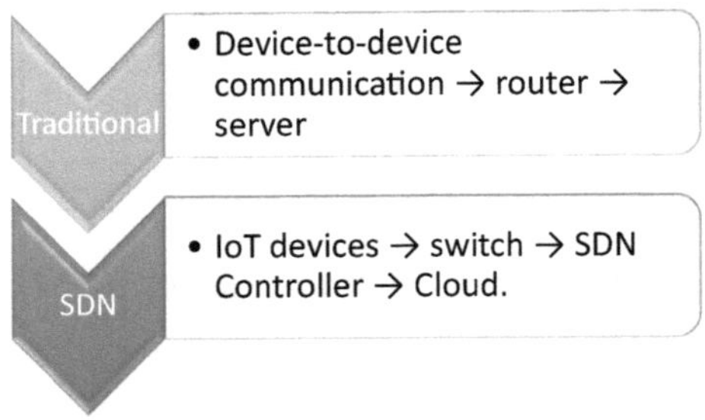

Fig. 7.2 A comparison between traditional network and SDN

7.2 Incorporation of SDN with IoT

Though the SDN was not developed keeping IoT in mind yet the inherent features of SDN are tailor made for the challenges IoT face. IoT realization is a challenging task/difficult problem as/because of the present IoT paradigm's generation of humongous data, intricate infrastructures, security flaws, and demands from recently created technology [13].

As discussed in previous section, the integration of SDN with IoT networks gives network capabilities of being scalable, flexible, programmable and centralized manged. The centralized control of SDN combined with the distributed aspect of IoT creates a robust foundation for efficiently managing complex and large-scale IoT systems. The SDN Controller is the core intelligence of the network while the gateway serves as the mediator between devices and controller. These capabilities lead to many benefits. Higher security, and improved resource management efficiency are two of them.

- IoT Devices: Different sensors, actuators, wearables and appliances are main IoT devices. These devices generate data and carry out operations. These devices are resource-constrained and therefore rely on gateway and controller for security and management tasks.
- Southbound Interfaces (SBI): These interfaces are responsible for communication between the SDN controller and IoT gateways/devices. Tasks like packet forwarding, device configuration are carried out utilizing protocols such as OpenFlow or NETCONF. SBIs help controller in enforcing decisions.

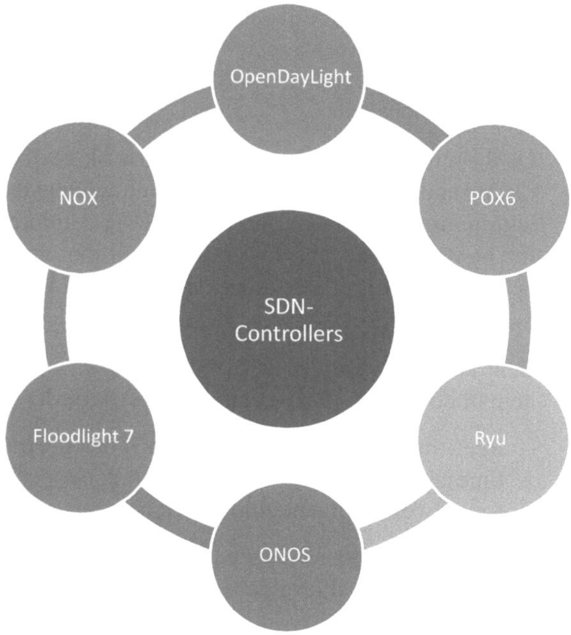

Fig. 7.3 Different SDN-controllers in-use [6–9]

Together, they form a powerful, efficient, secure, and scalable architecture.

The northbound interfaces (NBI) acts in 180 degree opposite direction to southbound APIs. NBIs provide channel of communication between IoT applications and the SDN controller. Through this, applications can request for network analytics, security and request for device configuration as per requirement. NBIs improve application intelligence, enable centralized policy control and enable automation (Fig. 7.3).

7.3 Advantages of Software-Defined Networking for IoT

1. Scalability and Flexibility

   Use of SDN makes a network scalable and flexible. These two characteristics provide many advantages to a network.

   - Simplified Network Expansion: As the size of organization grows so does number of devices and users in its network. SDN makes network scalable, allowing adding more devices, users, or services without redesigning the network. It enables quick expansion of network as organizations grow. It also helps in avoiding bottlenecks in performance or management.

- Dynamic Resource deployment: Scalable network SDN enables quick deployment of resources as the demand in network grows. As network grows, the demand for services also rises, a scalable network can support high-demand applications.

On the other hand, SDN makes the network flexible as well making it suitable for further and faster innovation. It also allows network to adopt newer technologies with minimal changes. Thereby, making it future-proof. SDN reduces demand for new hardware and lowers cost. As the changes are software based only, operational costs comes down and efficiency improves.

2. Improved Security—

   Centralized Security Policies: SDN follows a centralized architecture. With the help of controller, it can centrally apply security policies.

   - Real-time Threat Detection: Traffic flows are monitored at central level in real-time and can quickly isolate the malicious or compromised nodes. SDN allows for network segmentation which helps in controlling spread of malware.

3. Enhanced Network Management - SDN's programmability helps in configuring the nodes at central level, changing network settings and apply updates [14].

   - Automated Management: SDN routes traffic on real-time basis. It can prioritize the critical traffic and prevent congestions as well [14].

4. Quality of Service (QoS)—

   Traffic Prioritization: As mentioned in previous point, differentiation between critical and normal traffic helps in timely transmission of data of crucial data [10]. Utilizing the strategies focused on delay and loss of packets, one can harness the flexible and flow-based nature of SDN to improve QoS of the IoT applications significantly [15].

   - Bandwidth Management: Allocation of bandwidth based on demands of user and applications improves QoS and guarantees optimal performance [10]. SDN has been used for improving communication efficiency and QoS of IoT services [15].

7.4 Difficulties and Factors to Take into Account

SDN integration with IoT networks had benefits, it also poses several problems that must be resolved in order to fully use its potential.

7.4.1 Security Considerations

- Single Point of Failure: SDN is managed by an SDN controller which is single point of contact for all the services and communication. Therefore, its failure can become single point of failure. Ensuring the security and redundancy of it is of utmost importance.
- Data Privacy: The concentration of SDN might give rise to privacy problems, as sensitive IoT data is gathered and controlled in a centralized manner [16].
- Denial of Service (DoS) and Distributed Denial of Service (DDoS) can cause disruption of services. In these attacks, vulnerable nodes are targeted to create false traffic with false IP addresses to attack server. These attacks can be volume based, application layer focused, protocol layer focused and zero day. These can jeopardise the whole network given the centralized nature of SDNs [17]. It underlines the need to maintain a focused and proactive approach in innovation of new solutions to ensure the security of SDN from such attacks.

7.4.2 Challenges with Scalability

IoT networks may contain millions of devices, it results in enormous transmissions. To manage these transmissions SDN faces challenges due to its centralized architecture.

- Controller Performance: This humongous increase in transmissions results in more routing decisions, security checks, device registrations and real-time updates. It results in overloading of controller, slow-down of decision-making and lag of responses. This all affects functions of network and therefore, single controller may create bottleneck for entire network.
- Network Overheads: As discussed in previous point, all the rise in decisions, routings, checks and updates leads to increased network overhead. This impacts performance of network with higher latency, higher congestion, delayed actions and more processing load.

7.4.3 Interoperability and Standards

- Protocol Compatibility: IoT devices utilizes protocols such as MQTT, CoAP, Zigbee, Z-wave, LoRaWAN etc. while SDN controllers work upon OpenFlow, NETCONF, gRPC etc. These protocols operate at different layers and utilize varied data formats. To perform complex transaction, communicate properly, and apply uniform security policies without degrading network performance, it is necessary to achieve compatibility among these protocols [2–12].

- Standardization: The protocols, interfaces, management tools, and security mechanisms used in IoT systems are proprietary in nature, lacking standardization. It can lead to slow adoption, vendor lock-in, poor interoperability and security gaps.

7.4.4 Implementation Complexity

- Specialized Technical Knowledge: Integration of SDN-IoT network requires knowledge from the domains of SDN principles, network automation, virtualization, protocols, security and encryption etc. It presents a challenge for certain enterprises due to skill shortages, increased training costs, cross-disciplinary expertise.
- Transition from Legacy Systems: This transition to new technology has its own set of challenges such as hardware risks, compatibility issues, downtime risks, operational changes and high costs and resources.

7.5 Outlook and Patterns for the Future

The potential of SDN in IoT networks is promising, as continuous research and development efforts are focused on resolving existing obstacles and discovering novel opportunities.

7.5.1 Integration of Edge Computing

- Edge-SDN Convergence: Edge processes data on the device or any device connected to it. It will give the advantage of low latency, local autonomy and reduced bandwidth requirements. SDN can provide faster integration of edge devices, dynamic routing and bandwidth allocation and central oversight of network. This enhances the advantage of edge computing multiple time [18].
- Decentralized Control: The decentralized nature of edge computing all the challenges posed by the centralised behaviour of SDN controllers. It provides a decentralized SDN [18].

7.5.2 Integration of SDN in Different Future Networks

SDN provides remarkable efficiency in managing a network resources, SDN has been used for enhancing the resource management in optical networks as well making it better, smart and scalable [14, 19]. This capability, to manage massive number of devices running different services, makes SDN suitable option for creating flexible, scalable, cost-effective, and energy efficient future networks.

7.5.3 *Artificial Intelligence and Machine Learning*

- AI-Enhanced Management: Combination of AI and ML algorithms with SDN can help in providing intelligent network management, predictive maintenance, and adaptive security measures. These algorithms can enable SDN analyse network behaviour and take efficient decisions regarding routing, bandwidth allocation, and device configurations. It leads to better network management. By identifying patterns, this combination can predict link degradation and traffic anomalies in advance. It will timely intervention that will reduce downtime of network. Analysis of unusual traffic may result in prediction of attacks and an appropriate remedial action [10, 11, 19].

AI based failure recovery methods can help in quick detection of failures, intelligent re-routing of traffic resulting in minimization of service disruption. It will greatly improve the stability of data plane [4]. In addition to this, ML algorithms can help in traffic type classification, fast feature selection to help SDN in decision making [10].

- Improved Analytics: All these integration leads to important insights like, identification of bottlenecks, latency sources, or inefficient routes; detection of unusual traffic patterns, cyberattacks, or compromised IoT devices. These analytics can help in prediction and prevention of these problems through actionable intelligence.
- All these detections of problems and their remedies lead to self-healing capabilities. It guarantees networks stability and comprehensive delivery.
- The inclusion of AI-ML algorithms makes data plane intelligent and resilient. Fault tolerance is defined by the network being programmable which allows in dynamic updates. It should also allow the deployment of models. Final requirement is of standardizing northbound APIs so that controller and AI systems can interact easily.
- Federated learning trains models across and nodes without sharing raw data. It only shares updates. It helps SDN-IoT in reducing load by avoiding large data transfer. Privacy is enhanced as data remains on device. Learning is achieved from diverse environments. By optimizing routing decisions, bandwidth allocations and traffic scheduling it reduces jitter, packet loss, latency thereby improving QoS [19].

7.5.4 Standardization Initiatives

Standardization is necessary for interoperability, reliability, and large-scale deployment. Two

- Industry Collaboration: For developing unified standards, to harmonize protocols, to create secure and interoperable architectures and to accelerate adoption collaboration among industry stakeholders, standards bodies, and academics is of great importance.
- Open-Source Solutions: Open-source solutions allow for experimentation and innovation. The codebases are secure and transparent. It promotes broader involvement of community for improvement and development of standards.

7.5.5 Task Scheduling Approaches for Network Optimization

- Network tasks such as data processing, routing, allocation etc. are managed by task scheduling. These approaches determine how when and where these tasks will be performed. Through efficient task scheduling controller processing is optimized and unnecessary tasks are reduced. By preventing overloads and congestions, task scheduling improves network stability as well. In future, adaptive task scheduling algorithms needs to be developed so that it can change as per the demand. Traffic aware scheduling, predictive scheduling and edge scheduling are some of the algorithms that can be worked upon.

7.6 Summary

SDN offers optimum solutions for the diverse challenges faced by IoT networks. SDN integration with IoT system makes them scalable, flexible, secure and expanded network administration capabilities. Though many challenges need to be addressed, a continuous progress has been achieved in direction of making SDN based IoT networks better. In future, adaptive task scheduling, integration of AI-ML algorithms and edge computing to SDN will pave the way for development.

References

1. W. Jiang, Software defined satellite networks: a survey. Digit. Commun. Netw. **9**(6), 1243–1264 (2023). ISSN 2352-8648. https://doi.org/10.1016/j.dcan.2023.01.016
2. M. Awais, M. Asif, M.B. Ahmad, T. Mahmood, S. Munir, Comparative analysis of traditional and software defined networks, in *2021 Mohammad Ali Jinnah University International Conference on Computing (MAJICC)* (Karachi, Pakistan, 2021), pp. 1–6. https://doi.org/10.1109/MAJICC53071.2021.9526236.

3. M. Casado, M.J. Freedman, J. Pettit, J. Luo, N. McKeown, S. Shenker, Ethane: taking control of the enterprise. ACM SIGCOMM Comput. Commun. Rev. **37**(4), 1–12 (2007). https://doi.org/10.1145/1282427.1282382
4. N. Khan, R. Salleh, A. Koubaa, Z. Khan, M.K. Khan, I. Ali, Data plane failure and its recovery techniques in SDN: a systematic literature review. J. King Saud Univ. Comput. Inf. Sci. **35**(3), 176–201 (2023), ISSN 1319-1578. https://doi.org/10.1016/j.jksuci.2023.02.001
5. N. Gude, T. Koponen, J. Pettit, B. Pfaff, M. Casado, N. McKeown, S. Shenker, NOX: towards an operating system for networks. SIGCOMM Comput. Commun. Rev. **38**(3), 105–110 (2008). https://doi.org/10.1145/1384609.1384625
6. J. Medved, R. Varga, A. Tkacik, K. Gray, OpenDaylight: towards a model-driven SDN controller architecture, in *Proceeding of IEEE International Symposium on a World of Wireless, Mobile and Multimedia Networks 2014* (Sydney, NSW, Australia, 2014), pp. 1–6. https://doi.org/10.1109/WoWMoM.2014.6918985
7. F. Tomonori, Introduction to RYU SDN Framework (2013). https://ryu-sdn.org/slides/ONS2013-april-ryu-intro.pdf. Accessed 16 June 2025
8. P. Berde, M. Gerola, J. Hart, Y. Higuchi, M. Kobayashi, T. Koide, B. Lantz, B. O'Connor, P. Radoslavov, W. Snow, G. Parulkar, ONOS: towards an open, distributed SDN OS, in *Proceedings of the third workshop on Hot topics in software defined networking (HotSDN '14)* (Association for Computing Machinery, New York, NY, USA, 2014), pp. 1–6. https://doi.org/10.1145/2620728.2620744
9. S. Jain, A. Kumar et al., B4: experience with a globally-deployed software defined WAN. SIGCOMM Comput. Commun. Rev. **43**(4), 3–14 (2013). https://doi.org/10.1145/2534169.2486019
10. A.M. Eldhai, A. Abdelaziz et al., Improved feature selection and stream traffic classification based on machine learning in software-defined networks. IEEE Access, Digit. Object Ident. https://doi.org/10.1109/ACCESS.2024.3370435
11. F. Fernando Jurado-Lasso, L. Marchegiani et al., A survey on machine learning software-defined wireless sensor networks (ML-SDWSNs): current status and major challenges. IEEE Access Digit. Object Identif. https://doi.org/10.1109/ACCESS.2022.3153521
12. H.M.U. Rahman, R. Hanif, S. Ghafoor, A. Ahmad, Open Networking Engine (ONE): an orchestration tool for open optical line system. IEEE Access **12**, 8940–8956 (2024). https://doi.org/10.1109/ACCESS.2024.3354172
13. W. Rafique, A.S. Hafid, S. Cherkaoui, Complementing IoT services using software-defined information centric networks: a comprehensive survey. IEEE Internet of Things J. **9**(23), 23545–23569 (2022). https://doi.org/10.1109/JIOT.2022.3206146
14. C.N. Tadros, M.R.M. Rizk, B.M. Mokhtar, Software defined network-based management for enhanced 5G network services. IEEE Access **8**, 53997–54008 (2020). https://doi.org/10.1109/ACCESS.2020.2980392
15. N. Saha, S. BERA, S. Misra, Sway: traffic-Aware QoS routing in software-defined IoT. IEEE Trans. Emerg. Top. Comput. **9**(1), 390–401 (2021). https://doi.org/10.1109/TETC.2018.2847296
16. N. Anand, M.A. Saifulla, R. Babu Ponnuru, G. Reddy Alavalapati, R. Patan, A.H. Gandomi, Securing software defined networks: a comprehensive analysis of approaches, applications, and future strategies against DoS attacks. IEEE Access **13**, 64473–64515 (2025). https://doi.org/10.1109/ACCESS.2024.3520478
17. J. Polónio, J. Moura and R. Neto Marinheiro, On the road to proactive vulnerability analysis and mitigation leveraged by software defined networks: a systematic review. IEEE Access **12**, 98546–98566 (2024). https://doi.org/10.1109/ACCESS.2024.3429269
18. W. Rafique, L. Qi, I. Yaqoob, M. Imran, R.U. Rasool, W. Dou, Complementing IoT services through software defined networking and edge computing: a comprehensive survey. IEEE Commun. Surv. Tutor. **22**(3), 1761–1804 (2020). https://doi.org/10.1109/COMST.2020.2997475
19. A.F. Pakpahan, I.-S. Hwang, Peer-to-peer federated learning on software-defined optical access network. IEEE Access **12**, 84435–84451 (2024). https://doi.org/10.1109/ACCESS.2024.3411639

Chapter 8
Semantics and Its Importance in IoT

The concept of the IoT demonstrates a scenario that interconnects different objects and systems to allow them to share data and collaborate effortlessly. In order to realize this goal, it becomes necessary to establish a common understanding of the material that is being shared. It underlines the importance of the study of meaning in language. Semantics includes the analysis and comprehension of data to guaranty the uniform comprehension of the information throughout collaborating systems. This chapter studies the importance of semantics in the IoT, the difficulties faced in achieving semantic interoperability, and the strategies used in dealing with these obstacles.

8.1 Comprehending the Meaning of Semantics in IoT

The strategies used in robots for comprehension of data is serving as the base for various researchers to enhance the interpretation of data from various sources. Most common method is to integrate semantics into IoT data. The semantic web helps in better understanding of the environment by handling isolation issues among heterogeneous information. It is an extension of the World Wide Web. Pervasive computing is the realization of distributed computing which mandates the collaborative coordination across autonomous mobile hosts [1].

8.1.1 Semantics Definition

Semantics permits IoT devices to communicate, construe and comprehend data autonomously [1–3].

T. K. Sharma et al., *Understanding Internet of Things: Evolution, Enabling Technologies and Challenges*, Studies in Computational Intelligence 1282,
https://doi.org/10.1007/978-3-032-21588-8_8

Table 8.1 Key components of an ontology

Component	Description	Example (Smart Home Ontology)
Classes	Categories of objects	TemperatureSensor, SmartLight, HVAC System
Properties	Attributes of objects	hasTemperature, hasBrightness
Relationships	Connections between objects	SmartLight is ControlledBy SmartHub
Rules	Logical constraints	IF Room isOccupied THEN TurnOn SmartLight

– Semantics: The field of inquiry that investigates the meaning of language, encompassing the interpretation of words, phrases, and symbols. In the realm of IoT, data semantics is utilized to interpret the imortance and meaning of shared information instead of exchanging it like raw data.

8.1.2 Ontology

Ontology is a structured framework that presents knowledge as collection of concepts, relationships, and rules in purview of a particular domain. In IoT, semantic interoperability is enabled by ontologies. Ontologies ensures that disparate systems and devices can comprehend shared data in the appropriate context and with the appropriate meaning [2, 3] (Table 8.1).

In a smart healthcare system, an ontology would define:

- Patient, HeartRateMonitor, Doctor, and their relationships.
- hasHeartRate (HeartRateMonitor, Patient) to link a patient's IoT device readings.
- A rule like IF HeartRate > 120 THEN Alert Doctor.

Ontology-driven IoT allows systems to process data intelligently and automate decision-making.

8.1.3 Semantic Web Technologies for IoT

Linked data, reasoning engines, and established ontologies are utilized by the Semantic Web to improve the understanding of machine-readable data (Table 8.2).

In a smart hospital, monitoring devices, wearables, and medical devices are integrated with patients. Patient symptoms are interpreted contextually using RDF and OWL ontologies and physicians are notified of any unusual trends.

Table 8.2 Key semantic web technologies for IoT [1–3]

Technology	Role in IoT	Example
RDF (Resource Description Framework)	Represents IoT data in a structured, machine-readable format	A smart lightbulb stores its status (ON/OFF) as RDF triples
OWL (Web Ontology Language)	Defines relationships between IoT devices & actions	A "smart thermostat" ontology links temperature data to HVAC control rules
SPARQL (Semantic Query Language)	Extracts and processes IoT semantic data	Queries a smart building's RDF database for room occupancy patterns
SSN (Semantic Sensor Network Ontology)	Standardizes sensor data interpretation across IoT platforms	Smart agriculture sensors use SSN to analyze soil moisture data

8.2 The Significance of Semantics in the IoT

The field of semantics is crucial in multiple facets of the Internet of Things (IoT), enhancing the effectiveness, dependability, and expandability of IoT systems [4]. The Internet of Things (IoT) connects billions of devices, but these devices often:

i. Use different communication protocols (MQTT, CoAP, HTTP, Zigbee, Bluetooth, etc.).
ii. Generate diverse data formats (JSON, XML, CSV, proprietary formats).
iii. Lack a common understanding of data, making interoperability difficult.

Structure: Vertical stack of layers. From bottom to top:

1. Sensors is the bottom-most layer which collects raw data.
2. Processing layer performs filtering and analysis to refine the data so that further complex operations can be performed.
3. Semantic Layer is required to perform ontology mapping and reasoning on the refined output that has been received.
4. Service Layer then takes necessary action after intelligent decision-making.

For example, in a room a thermostat may provide a reading like Sensor reading = "25 °C", then its semantic interpretation will be "Comfortable Room Temperature" (Fig. 8.1).

Smart car parking is an important aspect of Smart City application. This parking will be made of different IoT sensors, traffic cameras for getting the real time update on the parking slots and mobile applications to identify and book these slots. These heterogeneous systems will their respective understanding about the 'vacant parking'. Without semantics, it can lead to confusion which will result in the inefficient performance of the system. This situation can be avoided by the use of semantic technology. Data models can utilize semantic technology to convey, understand and share data in a standardised way to create a seamless parking system. It will improve automation

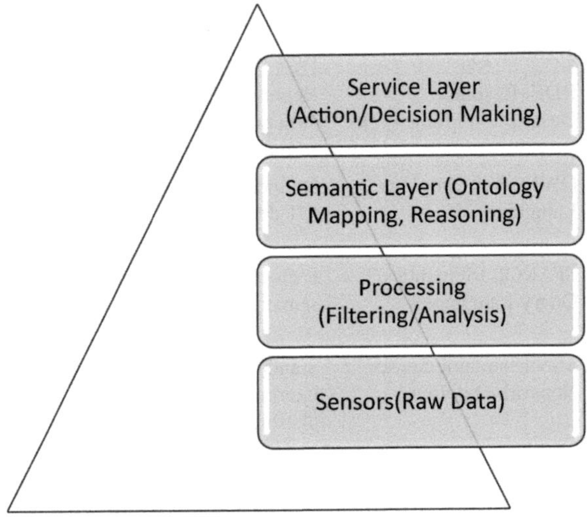

Fig. 8.1 Vertical stack of layers

and interoperability. Therefore, it can be concluded that lack of semantics leads to interoperability issues.

8.2.1 Sharing and Interpreting Data

Due to inherent heterogeneity in IoT, it is necessary that the information provided by one device must be read and used correctly. This can be guaranteed by combining and standardizing the data from various sources [4].

- Uniform Data Interpretation: The most important requirement for avoiding misunderstanding and errors in decision making is uniform interpretation of data. Let us suppose, a thermostat generates a data, "room temperature = 22 °C" and transmits it. Another device, a smart air-conditioner reads it then AC should interpret it as 22 °C and not 22 °F or 22% humidity. This type of misunderstanding can result in wrong decision. But if both devices agree on Semantic sensor network (SSN) ontology then both will understand 22 °C. In this way, Semantics ensures uniform understanding and analysis across devices.
- Contextual Understanding: Data does not mean much without context. In the Smart traffic management system of the Smart city provides the data "vehicle count = 50". This information in semantics along with context will provide a complete information with like "Location: Acacia Street, 5th Avenue"; "Time: Friday, 16:14", Vehicle Type = "private cars". With this context, smart traffic management system will understand this is rush hour congestion. The city's smart traffic lights and navigation apps can then it can extend green light duration on

Acacia Street. It can also redirect commuters to less busy routes. Therefore, semantics improves the interpretation of information as per the context which results in better decisions.

8.2.2 Improved Compatibility

IoT devices utilize "different languages" i.e. There are different data formats such as JSON, XML, CSV, proprietary codes etc. for representation. The communication methods are also diverse such as Wi-Fi, Zigbee, Bluetooth, LoRaWAN etc. These protocols have their own signal transmission rates. Finally, units/representations are also diverse for physical quantities (°C vs °F, "ON/OFF" vs "1/0"). This diversity in communication, representation and data formats makes device-cooperation a challenging task. Only a standards way to interpret and share data can help in this [5].

- Device Interoperability: The interoperability became a challenge when the devices are from different manufacturer. In a scenario of smart healthcare, if a patient is at his home-care and using multiple IoT health devices. Like, a heart-rate monitor from a Company Xand it reports "HR: 75 bpm." Maybe other blood pressure monitor from Company Y might report "BP: 120/80 mmHg." Any other device from blood glucose sensor from company Z reports "*Glucose: 110 mg/dL.*" It will make it difficult for the smart health application to combine them. But with semantics, all devices will be able to map their outputs to a common ontology (HL7 or SNOMED CT). It will combine "HR," "Pulse," and "BP" and hospital dashboard can integrate and display these parameters seamlessly. In this way, semantics can help doctor to get one unified report of the patient's health, independently of device type, brand or data format.
- System Integration: In smart city concept, many systems do their operate at tandem like traffic sensors to detect congestion, air-quality sensors to measure pollution etc. Some public transport applications may also be used to locate bus with GPS and identify passenger load. Integration of these systems will be difficult with each other as public transport information will be in logs, air-quality data in XML and traffic data might be in JSON format. Utilizing semantics, e.g. SSN ontology or or FIWARE NGSI, all these parameters can be expressed in a machine understandable and uniform way. It can help smart city application to minimize the private vehicle and increase bus frequency in case of high air pollution situation. Application will also send alerts to citizens regarding pollution situation. Thereby, countering a problem with optimum decision making.

8.2.3 Enhanced Decision-Making and Automation

As discussed previously, IoT applications are complex system. These can derive at conclusions by utilizing the data from multiple sources. Therefore, context-aware data-fusion, data interpretation and extraction of knowledge from this refined data are of utmost importance.

- Precise Data Analysis: Semantic reasoning not only helps in the fusion of data from multiple sources but also makes it rich contextually. It also standardizes the fused data for consistent understanding. This helps in avoiding the misconstrued inferences and providing accurate intelligence. In smart agriculture scenario, soil-moisture sensors, weather station and crop growth sensors may provide their respective inputs as 25%, 0 mm rainfall, plant health → moderate. This system may misinterpret the information as the "enough water" without linking it to crop type or weather (Fig. 8.2).

But, inclusion of semantics changes it increases the accuracy of data analysis by providing meaningful and contextually thorough data. E.g. in the above scenario semantic inclusion will help understand that the '*crop is rice*' i.e. it requires high moisture. This will change '*0 mm rainfall*' to be interpreted as '*need for irrigation*'. Additionally, '*plant health = moderate*', will raise a warning. Hence, system will respond with a precise suggestion of "*Irrigate for 2 h for achieving optimal moisture content in soil.*"

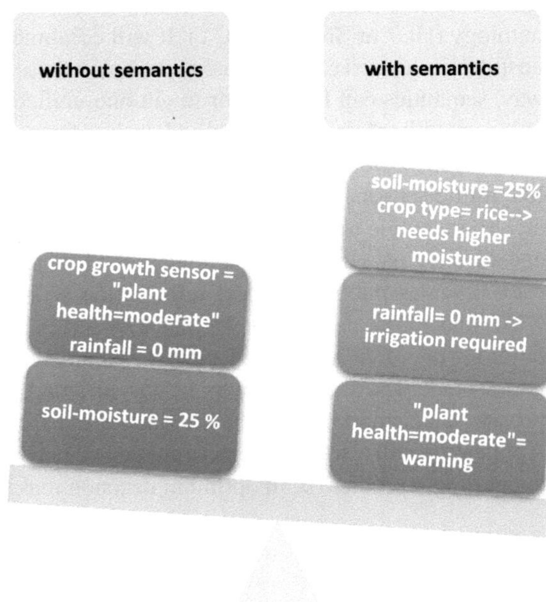

Fig. 8.2 Semantic effects on precise data analytics

- Real-Time Insights: Semantics reasoning will also help in the real time interpretation and decision-making especially for the applications like smart traffic management where response time should be minimum. Let us analyse one such case where CCTV based monitoring may result in "*vehicle density = high on 3rd street*", Air Quality sensors respond "*PM 2.5 = 120 (unhealthy)*" and public transport GPS may reflect "*Next bus arriving in 3 min*". In absence of semantics, system might process these three datasets separately. With semantics, system will be able to link PM2.5 concentration with high traffic. Understanding that people may switch their mode of transport it may extend green signal to bus and send the alerts to the people "*High pollution and traffic congestion ahead, you're suggested to take bus and not drive.*"

8.3 Difficulties of Achieving Semantic Interoperability in the Internet of Things (IoT)

Every technology has to face various challenges during its development. In case of Internet of Things (IoT), ensuring the correct understanding and utilization of communicated data among various devices and systems is major challenge. It comes under the challenge posed for semantic interoperability.

8.3.1 Device Variability

- Heterogeneous Standards: IoT systems utilizes devices from different vendors which uses diverse communication protocols, data formats, and standards. It poses difficulties in attaining semantic interoperability. For example, in smart agriculture system, Soil sensor using MQTT protocol will send data {"soil_moisture": 45} in json format. While soil sensor using CoAP protocol communicates <moisture> 0.45 </moisture> in php template. It is a challenge for Gateway that data should be universally as 45%. Heterogeneity in protocols make it difficult to connect devices. On the other hand, various data formats make it very hard to parse of information.
- Diverse Capabilities: Processing power, memory, and connectivity of the IoT devices also pose a challenge by making it difficult to create consistent semantic frameworks.

8.3.2 *Absence of Shared Ontologies*

- Ontology Definition: *Ontologies are the formal and structured description of entities, their properties and relationships with environment and other entities.* It helps in uniform understanding in the context of the IoT.
- Variety of Domains: As the domain changes, such smart homes, healthcare, and industrial IoT, its related entities, their properties and relationships also change. These changes make the development of a universal ontology a tough challenge.

8.3.3 *Data Context and Semantics*

- Contextual Variations: Context is very important in IoT applications. It represents the location, time, environment, user activity, or purpose. Semantic meaning of a data depends upon '*where and how*" the data was generated. With different contextual information, same data can provide different interpretations and lead to different decisions. Therefore, it is very difficult to maintain a consistent or "uniform" interpretation across various applications.
- Dynamic settings: IoT settings are highly dynamic. Sensors can join or leave the network very frequently. Similarly, locations may also change and capabilities such as firmware or data-types may also be updated. Any such change necessitates the reconfiguration of ontology mappings and contextual information. It also requires a real-time semantic discovery and adaptation (Fig. 8.3).

8.4 Addressing Semantic Interoperability Challenges

IoT systems face different semantic interoperability challenges. Many approaches and solutions have been provided by the researchers to address these challenges.

8.4.1 *Creation of Uniform Ontologies*

- Industry Standards: Lack of available industry standards for IoT devices is one challenge that creates problem. For example, a smart thermostat may report "temp = 22 °C". On the other hand, a smart AC from different brand may only understand "temperature = 295 K." Any other device from separate vendor might use "RoomHeat = 22". In this scenario, though these all mean same but system may not interpret correctly without standardized structure. Hence, interoperability becomes an issue.
- Industry-wide ontologies will create consistency independent of vendors and provide consistency in the meanings. One such example is W3C Semantic Sensor

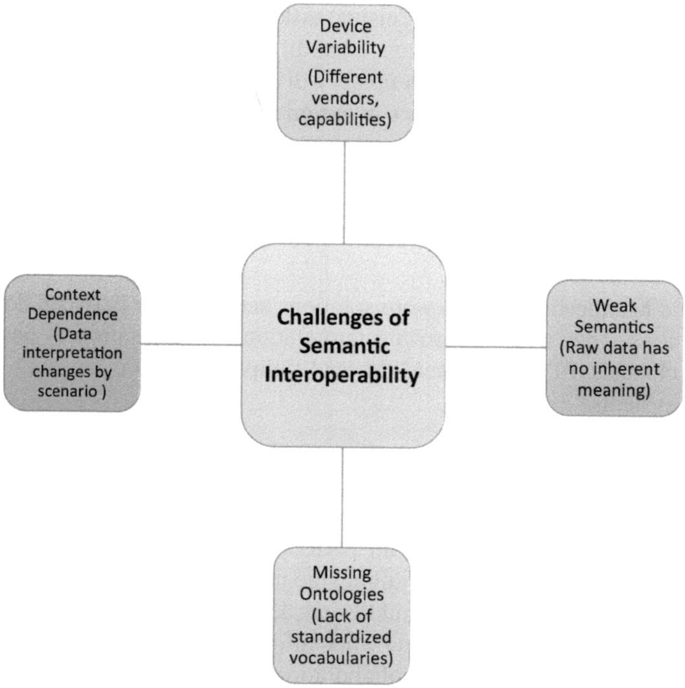

Fig. 8.3 Challenges with semantic interoperability

Network (SSN) ontology. It offers a standardized structure for defining sensors, their observations, and related data. Thereby, promoting consistency among different systems. Suppose, in a smart city a deployment of IoT systems from three different vendors to monitor air quality is made. The following observations might look like this:

1. *Vendor X's sensor output: {"temp": 22, "pm25": 40}*
2. *Vendor Y's sensor output: {"temperature": 295 K, "fineDust": 40 μg/m*^{3}*}*
3. *Vendor Z's sensor output: {"RoomHeat": "22C", "PollutionLevel": 40}*

Now, challenge arise from the terminology used in these devices. Labels "temp," "temperature," and "RoomHeat" are not identified as the same concept by all these systems. Different units, Celsius vs Kelvin, might not be identified. Labels such as "pm25," "fineDust," and "PollutionLevel" though these all mean PM2.5 concentration but are mentioned in a different way.

In case of semantics, every sensor maps its data into SSN standard form:

1. Temperature → ssn:Temperature
2. PM2.5 → ssn:Concentration with unit μg/m^3

The unit conversion (Kelvin → Celsius) task is carried out by ontologies. Now, dashboard can integrate these readings from different vendors without any hinderance. In this way, a *'unified air quality index'* will be created for small city application. With this application, the authorities can optimise their decisions based on these.

8.4.2 Implementation of Middleware Solutions

- Semantic Middleware: Middleware solutions act as bridge between devices and applications. They help in correct interpretation and adaptation as per environmental changes. FIWARE, oneM2M, and Eclipse Kura, are few such solutions for IoT systems.

Smart Appliances REFerence Ontology (SAREF) is an ontology developed by European Telecommunications Standards Institute (ETSI). It defines shared vocabulary for devices functions, services and properties in smart city, energy and buildings etc. Devices, who complies with SAREF, provide their functions, units of measurements and relationships on joining the network.

FIWARE is an open-source IoT middleware designed to track real time changes in IoT environment (i.e. what, where and when information regarding changes). In this way SAREF and FIWARE works in collaboration to ensure semantic interoperability.

8.4.3 Utilization of Machine Learning and Artificial Intelligence Techniques

When data from multiple sources is utilized by IoT systems, these may face problems due to data formats, naming conventions and ontologies. Semantic mapping and AI systems helps in sorting these problems.

- Semantic Mapping: It helps in maps data from many sources to a common meaning. As labels are used in ontologies. AI systems are capable of analyse and detect patterns in data labels, values and relationships. These systems can also learn mapping. These will align datasets to a common ontology. These capabilities help these systems to adapt to dynamic behaviour of IoT systems. In this way if these two are used in conjunction they will create seamless data sharing [6, 7].

8.4.4 Context-Aware Systems

- Dynamic Context Handling: As discussed previously, that context is quite important in exact interpretation of the information. Therefore, context aware systems ensure understanding of the information. They provide semantic consistency in dynamic environments. Thereby, establishing reliable automation through seamless data exchange [8].

8.5 Case Studies: Resolving Semantic Interoperability Challenges in Smart Cities

The heterogeneity of devices, their allocation, and the "local-to-global" relationship makes the IoT programming a complex task. It requires addressing following three dimensions:

(i) the management of property perception from the IoT environment via sensors (sensing dimension),
(ii) the processing of sensor information to make decisions for fulfilling system objectives (processing dimension), and
(iii) the implementation of decisions through actuators (actuation dimension).

To understand this more comprehensively a case study from smart cities has been discussed here:

8.5.1 The Difficulty

In a smart city, various IoT devices and systems from multiple vendors collaborate to deliver services such as traffic control, waste management, and public safety etc. It must attain semantic interoperability to guarantee effective exchange and comprehension of data.

8.5.2 The Resolution

- The city can implement standardized ontologies, such as the SSN ontology, to achieve consistent description of sensors and their data.
- Semantic Middleware: Further, a semantic middleware platform can be used to convert heterogeneous data into a unified semantic representation.
- AI-Powered Semantic Mapping: Finally, AI techniques can be employed to autonomously map and merge diverse data for improving coherence and usability (Fig. 8.4).

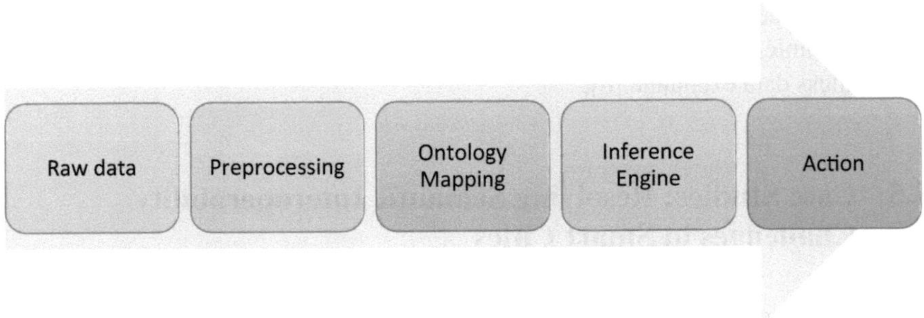

Fig. 8.4 Flow with semantic interpretation

8.5.3 Result

By focusing on semantic interoperability, the smart city was able to:

- Optimize Service Efficiency: Enhanced the efficiency of city services by integrating and analyzing accurate and real-time data.
- Promote Interoperability: Fostered collaboration among diverse systems and divisions, augmenting overall functionality and performance.
- Enhance Citizen Experience: Implemented improved services and responses to citizens by utilizing increased data-driven decision-making.

8.5.4 Semantic IoT in a Smart Home

Scenario:

Imagine a smart home where different devices like thermostat, lights, security cameras, door locks etc. must work together. Without Semantics the following issue may arise:

1. A motion sensor detects movement and sends { "motion": true} in JSON format.
2. The security camera logs movement but uses a different XML format: <event> <motion> true </motion> </event>
3. The smart door lock system does not recognize either of these formats, causing a delay in responding to a potential break-in.

Hence, the problem arises that different devices cannot interpret each other's data.

8.5.5 *Case Study: Semantic Interoperability in Smart Homes*

Problem: Lack of IoT Interoperability in Smart Homes

Smart homes consist of multiple IoT ecosystems such as Google Home, Alexa, SmartThings, Apple HomeKit etc. Devices from different vendors do not share a common data model, preventing seamless automation. A solution may be derived using semantic ontologies data linked data.

Develop a Smart Home Ontology defining:

- SmartDevice, Sensor, Actuator, AutomationRule.
  Use Linked Data to connect different IoT platforms.
  Build a Knowledge Graph that links all smart home devices and their behaviors.

Example Ontology Rule in a Smart Home:
iot:MotionSensor iot:triggers iot:SmartLight.
iot:SmartLight iot:hasStatus "ON".

Now, motion detection automatically turns on the light, regardless of device vendor.

Finally, Fig. 8.5 illustrates some significant ontologies that are accessible from the viewpoint of the reader. The SSN ontology was created in 2011 by the SSNXG, the W3C semantic sensor network incubator group. The sensor's capabilities, operations, measurement, etc. are represented by ontologies like CSIRO and the Coastal Environment Sensor Network (CESN). The SSN ontology ideas have been extended to generate the IoT-A and IoT.est models. Additionally, it introduced several IoT concepts, like services and items connected to sensor devices [9–19].

8.6 Summary

The success of IoT relies heavily on semantics, which allows for uniform interpretation of data, greater interoperability, and enhanced decision-making. Although semantic interoperability is difficult due to the diversity of devices, absence of shared ontologies, and contextual variations, several methods such as standardized ontologies, semantic middleware, artificial intelligence techniques, and context-aware systems can be employed to address these challenges. By focusing on semantic interoperability, IoT systems can achieve smooth and efficient communication, integration, and functionality, leading to innovation and increased efficiency in a wide range of applications.

Fig. 8.5 Different ontologies available in literature [9–19]

References

1. Y. Huang, G. Li, A semantic analysis for Internet of Things, in *2010 International Conference on Intelligent Computation Technology and Automation* (Changsha, China, 2010), pp. 336–339. https://doi.org/10.1109/ICICTA.2010.73
2. I. Ullah, D. Adhikari, X. Su, F. Palmieri, C. Wu, C. Choi, Integration of data science with the intelligent IoT (IIoT): current challenges and future perspectives. Digit. Commun. Netw. **11**(2), 280–298 (2025). ISSN 2352-8648. https://doi.org/10.1016/j.dcan.2024.02.007
3. A. Rhayem, M. B. A. Mhiri, F. Gargouri, Semantic web technologies for the Internet of Things: systematic literature review. Internet of Things **11**, 100206 (2020). ISSN 2542-6605. https://doi.org/10.1016/j.iot.2020.100206
4. S.K. Gupta, R.R. Chandan, R. Shukla, P. Singh, A.K. Pandey, A.K. Jaiswal, Heterogeneity issues in IoT-driven devices and services. J. Auton. Intell. **6**, 13 (2023). https://doi.org/10.32629/jai.v6i2.588
5. M. Ganzha, M. Paprzycki, W. Pawłowski, P. Szmeja, K. Wasielewska, Semantic interoperability in the Internet of Things: An overview from the INTER-IoT perspective. J. Netw. Comput. Appl. **81**, 111–124 (2017). ISSN 1084-8045. https://doi.org/10.1016/j.jnca.2016.08.007

6. L. Waltersdorfer, A. Breit, F.J. Ekaputra, M. Sabou, Bridging semantic web and machine learning: first results of a systematic mapping study, in *Database and Expert Systems Applications—DEXA 2021 Workshops. DEXA 2021. Communications in Computer and Information Science*, vol. 1479 (Springer, Cham, 2021). https://doi.org/10.1007/978-3-030-87101-7_9
7. P.K. Agrawal, A novel mapper machine learning algorithm for semantic domain mapping for domain database updation. SN Comput. Sci. **4**, 536 (2023). https://doi.org/10.1007/s42979-023-02036-0
8. T.Y. Lai, I.C. Hsu, Integrating semantic web into context-aware mobile application based on cloud computing. J. Cloud. Comp. **14**, 8 (2025). https://doi.org/10.1186/s13677-025-00731-2
9. K. Pal, A.U.H. Yasar, Semantic approach to data integration for an Internet of Things supporting apparel supply chain management. Procedia Comput. Sci. **175**, 197–204 (2020). ISSN 1877–0509, https://doi.org/10.1016/j.procs.2020.07.030
10. M. Compton, P. Barnaghi, L. Bermudez et.al., The SSN ontology of the W3C semantic sensor network incubator group. J. Web Semant. **17**, 25–32 (2012). ISSN 1570-8268. https://doi.org/10.1016/j.websem.2012.05.003
11. S. Cox, Ontology for observations and sampling features, with alignments to existing models. Semant. Web J. **8**(3), 453–470 (2017). https://doi.org/10.3233/SW-160214
12. W. Wang, S. De, R. Toenjes, E. Reetz, K. Moessner, A comprehensive ontology for knowledge representation in the Internet of Things, in *IEEE 11th International Conference on Trust, Security and Privacy in Computing and Communications* (Liverpool, 2012), pp. 1793–1798. https://doi.org/10.1109/TrustCom.2012.20
13. M. Aziez, S. Benharzallah, H. Bennoui, An ontology-based context model for the discovery of IoT services in the Internet of Things, in *International Conference on Mathematics and Information Technology (ICMIT)* (Adrar, 2017), pp. 209–213. https://doi.org/10.1109/MATHIT.2017.8259719
14. M. Bermudez-Edo, T. Elsaleh, P. Barnaghi, K. Taylor, IoT-Lite: a lightweight semantic model for the internet of things and its use with dynamic semantics. Pers. Ubiquit. Comput. **21**, 475–487 (2017). https://doi.org/10.1007/s00779-017-1010-8
15. N. Seydoux, K. Drira, N. Hernandez, T. Monteil, IoT-O, a core-domain IoT ontology to represent connected devices networks. Eur. Knowl. Acquisition Workshop **561–576**(2016). https://doi.org/10.1007/978-3-319-49004-5_36
16. A.K. Mandal, A. Cortesi, A. Sarkar, N. Chaki, Things as a service: service model for IoT, in *IEEE 17th International Conference on Industrial Informatics (INDIN)* (Helsinki, Finland, 2019), pp. 1364–1369. https://doi.org/10.1109/INDIN41052.2019.8972241
17. S.K. Mishra, A. Sarkar, Service-oriented architecture for Internet of Things: a semantic approach. J. King Saud Univ. Comput. Inf. Sci. **34**(10), 8765–8776. ISSN 1319-1578 (2022). https://doi.org/10.1016/j.jksuci.2021.09.024
18. H. Rahman, I. Hussain, A comprehensive survey on semantic interoperability for Internet of Things: State-of-the-art and research challenges **31**(12), e3902 (2020). Special Issue: "Cross-layer innovations in Internet of Things" and "Advanced Microprocessor Optimization methods for the Internet of Things". https://doi.org/10.1002/ett.3902
19. L. Daniele, F. den Hartog, J. Roes, Created in close interaction with the industry: the smart appliances REFerence (SAREF) ontology, in *Formal Ontologies Meet Industry. FOMI 2015. Lecture Notes in Business Information Processing*, vol. 225, ed. by R. Cuel, R. Young (Springer, Cham, 2015). https://doi.org/10.1007/978-3-319-21545-7_9

Chapter 9
Use Cases and Market Opportunities

9.1 Introduction

IoT offers immense possibilities of Business for all the stakeholders, that can help develop a Use Case into a service. Various technologies are required to deliver a service in IoT from WSNs to Big Data techniques. From cellular operators to cloud service providers, everyone will get their share of market capitalization. IoT is a dream that has yet to realize its full potential. There are numerous studies by major firms like Mckinsey and Company which gives a detailed account of market potential of IoT. According to [1], the number of connected devices would reach 39 billion by 2030. It puts the current number of connected devices at 21.1 billion. The data presented here simply points to the fact that device industry will see a continuous growth in the in number of device requirement.

Business generated by the IoT use cases can affect the market in two ways. First, it can directly affect the market in terms of business generated. Second, it will have economic impact on the businesses in terms of reduction of costs, increased productivity etc. There is variety of stakeholders that are part of this like service providers, research centers, network operators and development organizations etc. RFID products and services are growing at a very fast rate. With a CAGR of 9.1%, it will reach 25.24 billion in 2033. Currently, it stands at USD 12.61 billion in 2025 [2].

Industrial Internet of Things (IIoT) would be the biggest contributor to the market. As per Market&Markets, $ 198.2 billion in 2025 to $ 286.3 billion in 2029. IIoT is expected to grow at CAGR of 8.1% [3]. Grand View research [4] mentions that the market size of m-health products which stands at $ 62.7 billion in 2024 will reach at $158.3 billion by 2030 with a CAGR of 14.1%. The main reason behind this growth will be smartphone penetration. The major contributor to this market will be exercise and nutrition applications which will contribute at 34 and 22% of the total. From user point of view, It also says that M-health applications for monitoring and Managing Illness, Improving Wellness, will 3rd major contributor in the market capitalization of IoT while the first two will be factories and Cities.

T. K. Sharma et al., *Understanding Internet of Things: Evolution, Enabling Technologies and Challenges*, Studies in Computational Intelligence 1282,
https://doi.org/10.1007/978-3-032-21588-8_9

In light of the presented data, one can say that IoT will be a thriving market in the time to come which will impact all aspects of life. This provides the researchers; business organizations and Internet service providers with the opportunities are a part of this "wave of Smartness".

Services provided by these smart objects form two types of markets, Horizontal markets and Vertical markets, and these markets have potential of billions and trillions of US Dollars as projected by various research. Figure 9.1 illustrates the interaction between services in IoT. The experts of the IoT markets talk about two types of service markets:

i. Vertical Service Markets (Verticals)
ii. Horizontal Service Markets (Horizontals)

The services that belong to a particular domain comes under verticals while the services that can be utilized in any application are termed as horizontals. Normally, Vertical Markets are focused upon a single type of service. If all the services for a use case are provided by a single company, it would be considered as Vertical e.g. a smart home service where sensing to decision making are provided by a single company. On the other hand, if 'sensing' is provided by one company, 'connectivity' by another and third company which utilizes these services to deliver 'Smart Home' service then it should be considered as a case of horizontal services.

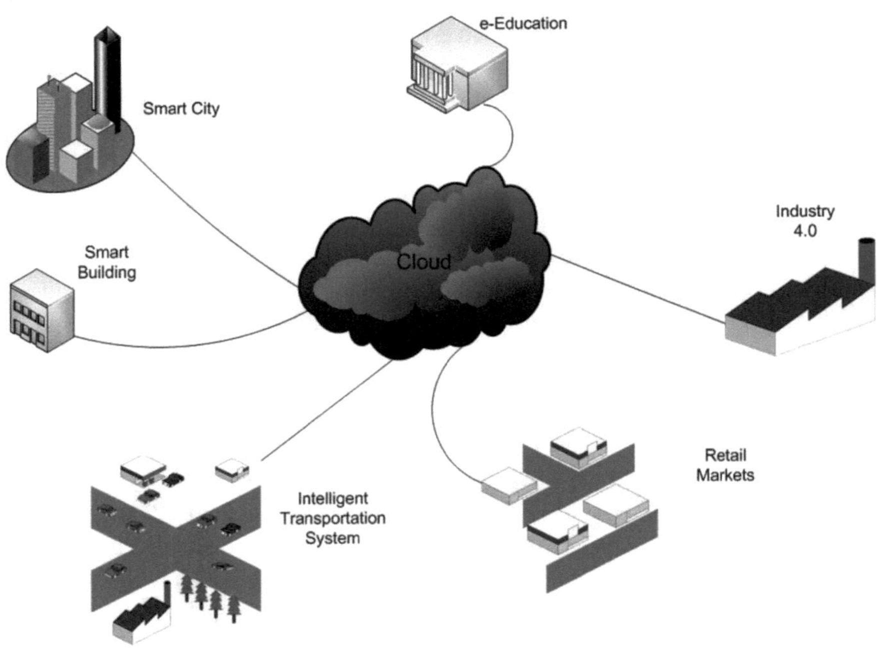

Fig. 9.1 Internet of things use cases

9.2 Use Cases

IoT is a technological evolution that connects ordinary devices to the internet and allows them to exchange data. This interconnection enables unparalleled levels of automation, efficacy, and data-driven decision-making across diverse industries. The global Internet of Things (IoT) industry has experienced swift expansion, propelled by technical progress, decreasing costs of sensors and connectivity, and the growing need for intelligent and interconnected devices. This chapter examines the many applications of IoT in different sectors and investigates the market opportunities that result from these uses.

9.2.1 Smart City

'Smart City', which is combination of two terms one is 'city' signifying a large densely populated geographical area with all the necessary means to support this population in terms of infra, health, transportation, communication, government and energy. While 'smart' signifies inclusion of diverse technologies, systems and policies that help optimize the resource consumption to deliver benefits to inhabitants [5].

Upgradation of this sector relied mainly on one of the following four key developments.

1. Inherently low-cost and real-time monitoring capabilities of IoT systems makes them suitable for interlinking different parameters of smart city to each other. In addition to these, low latency and mobility make these systems highly preferrable.
2. Cyber-Physical-Systems (CPS) can perform various functions such as automated driving, cobots etc. In these systems, AI and ML is used along with a combination of multiple systems for carrying out the specialized operations that only humans could have performed till now.
3. Enormous amount of information collection, its storage and processing, lies beneath the operations of smart city. The modern multifunctional devices require less maintenance, and consumes very small power. These modern devices deliver services on user's requests with few or no errors by taking real time-decisions.
4. The relationship between users and devices is the fourth most important pillar in this context. It comes under management strategies which designs all the directives, regulations, laws, or policies to direct and control device-user relationships. As user and device are integral part of this system, each one is ignorant of the complete characteristics and behaviors of the other. The main objective is to develop these strategies in a way to bring out the positive points and avoid the points which can put them on disadvantage.

9.2.2 Smart Homes

Smart houses are widely recognized as one of the most prominent implementations of IoT. These homes integrate internet connectivity to enable remote management of various equipment, including thermostats, lighting systems, security cameras, and home appliances, through smartphones or voice assistants such as Amazon Alexa or Google Assistant. The capacity to mechanize and regulate domestic surroundings has resulted in substantial convenience, energy conservation, and heightened security.

Utilization Scenarios

- Smart Lighting: In smart homes, one important way to ensure energy/ power savings is to control lights. Smart lightning systems which can be controlled by mobile applications plays important role. Phillips Hue and LIFX are example of such systems which can be controlled remotely through apps. If these smart lightning systems can be connected with motion sensors, then these can stop the unnecessary wastage of energy.
- Smart Thermostats: Machine learning algorithms infused inside IoT devices can help learn the personal preferences of occupant of a house. It can be utilized to create more personalized experience for the inhabitants. Google Nest thermostats, adjust the home temperatures based on the occupancy and saves he electricity.
- Home Security Systems: Different IoT devices can work in collaboration to ensure the safety of smart house. Use of smart cameras, door-bells and smart locks can be entwined to ensure the safety of inhabitants of the house. Notifications can be sent to the homeowners in case of any abnormal activity. It will lead to prompt and necessary action required for the safety.

Market Opportunities

- The smart home market is set to experience significant expansion, fueled by the rising consumer desire for convenience, security, and energy efficiency. Statista's analysis predicts that the worldwide smart home industry will have a value of $174 billion by 2025. With a CAGR of 9.55%, it is expected to reach US $250.6 billion by 2029. The main components of this market are smart appliances, comfort and lightning, control & connectivity, security, home entertainment and energy management [6].
- Home Automation: This segment actually belongs to smart appliances which are at the helm of all the automation. Their combination provides a potential commercial aspect. Highly reliable and user-friendly platforms from companies will have an advantage.
- Energy Management: Most of the sustainability requirements arise from the intelligent use of energy. It costs a lot if the people spend energy recklessly. Hence, the appeal of the companies is growing day by day that provide solutions for maximum utilization of energy. These products save a lot of money for consumers.
- Security Solutions: The growing apprehension regarding home security offers a substantial market potential for security devices based on the IoT. Companies

who are able to offer cost-effective, dependable, and intuitive security solutions will dominate a significant portion of this market.

9.2.3 Healthcare Industry

Internet of Medical Things (IoMT) is the term coined for the usage of IoT devices in the healthcare industry. IoMT applications and systems can be deployed for monitoring the patients, making predictive analysis from the data collected by the smart devices. It has transformed the patient care to preventive care and improved the efficiency in healthcare delivery.

Utilization Scenarios and Market Opportunities

- Remote Patient Monitoring (RPM): Patients are observed by doctors and nurses through their vitals. The important parameters include oxygen levels, blood pressure, glucose levels, heart rate and body temperature. This data can be collected via smart wearable IoT gadgets. Reviewing and analysing data medical practitioners can provide medical interventions at appropriate times. It can avoid unnecessary visits to doctors and help elderly get the hassle free medical-care. Thereby, saving time, money and efforts of both patients and practitioners.
- Smart Medical Devices: Integration of technology, connectivity and AI in healthcare devices has allowed them to become intelligent. It has allowed them to collect, analyse and transmit the vital information from patients. Now, they not only record the data but also connect with other devices to send alerts or analyse the patterns. Smart inhalers, Smart watches, smart pacemakers, remote ECG patches are just a few examples of such devices. These types of devices can facilitate early diagnosis, personalized treatment and continuous monitoring.
- Predictive Maintenance of Medical Equipment: IoT sensors combined with data analytics and AI can be utilized to predict failure or maintenance requirement of a healthcare device. It helps in minimizing the downtime of the equipment, prevent its malfunctions and minimizes repair costs. It will ensure that there is no disruption of patient care.

Market Opportunities

The IoMT industry is flourishing day by day. The increasing use of smart medical equipment has led to tremendous growth in this sector. The two important factors behind this surge are improvements in wearable technology, and the rising demand for remote healthcare services. As per Grand View Research, the market for IoMT is estimated at the value of $230.69 billion in 2024. It will grow at a CAGR of 18.2% and is projected to reach at US $658.57 billion by 2030. North America being the largest market in the 2024. Currently hardware has the largest market share which will be overtook by the Software in 2030 [7].

- Telemedicine has the biggest share in the current market share. It employs the telecommunication technology in the remote delivery of healthcare services. It will be instrumental in achieving the United Nations Sustainable Development Goal by providing health services for all. The second largest share goes to clinical operations and workflow management. Companies like Apple, Boston Scientific and CISCO etc. are the main contributors who have the ability to provide precise, dependable, and easy-to-use wearable gadgets will attract a significant portion of this market.

– Remote Patient Monitoring Services: Utilizing remote monitoring systems can help identify early warning indicators of health issues and allow for prompt intervention by gathering information on a patient's condition, such as blood pressure, glucose levels, and heart rate, and sending it to medical professionals for assessment and observation. These services can help in improving medical results for patients. Chronic conditions can be managed effectively allow for timely medical intervention when required. It cuts down unnecessary visit to doctor, thereby, cutting down healthcare costs. There is large scope for such companies who can offer such services.
– IoMT-based RPM framework when integrated with AI-ML algorithms will optimize the human involvement in many healthcare procedures such as disease diagnosis, medical picture interpretation, treatment planning etc. These algorithms have huge potential to identify and understand the underlying and hidden patterns in extensive patient data. Use of AI-ML algorithms perform more precise diagnoses and formulates personalized treatment programs to improve the efficacy of treatment.

9.2.4 Industrial Applications

IIoT paradigm uses IoT devices and techniques for production and automation. IIoT improves productivity, reduces downtime, and enhances safety through real-time surveillance and predictive maintenance from sensor feeds. Its market stands at $198.2 bn in 2025 that will grow at a CAGR of 8.1% to reach $ 286.3 bn in 2029 [3].

Utilization Scenarios

– Predictive Maintenance: Different IoT sensors are deployed on machines to collect temperature, vibration, noise levels, pressure, and other performance metrics. These real-time parameters are used as input to AI-ML algorithms to predict the possible breakdowns or faults in advance. Advance prediction of unexpected failures, can help in schedule maintenance so that productions does not interrupt for long. It also avoids mishaps and saves life and money.
– Supply Chain Management: IoT systems with help of GPS and RFID devices facilitate continuous tracking of items and resources across the globe. It helps in identifying demand, managing inventory, reducing delays in replenishment.
– Automation and Control: IIoT systems help in adjusting production parameters, and monitoring environmental conditions to ensure production quality. This information can be used resulting in enhanced efficiency, consistency, and safety. It

enhances automation by improving the methods of machine handling. It also adjusts the parameters of production such as flow rates, machine input and speed, thresholds for quality testing etc. In this way, the safety and efficiency are improved.
- Predictive Maintenance Solutions: Organizations that are into business of providing predictive maintenance solutions such as analytics software, sensors, and monitoring platforms, have significant business potential. These solutions can be used to reduce maintenance costs, improve operational efficiency, extend asset life and carry out safer operations.
- Real-Time Analytics and Data Management: converting the data from machines and IIOT sensors into actionable insights can lead to operational efficiency, improved productivity and better resource allocation. Companies offering such data platforms, edge analytics solutions, cloud base monitoring systems can derive business from the industries that require continuous operational insights.
- Automation and Robotics: Demand for higher automation, labour shortages, precise robotic operations, growth of industry 4.0 has generated huge demand for this sector.

9.2.5 *IoT in Urban Areas*

Multiple IoT technologies are utilized in smart cities for the enhancement of the well-being of its citizens. Optimum utilization of urban infrastructure, efficient delivery of public services, and acting on the environmental conditions are the main objectives that are targeted by these systems. Intelligent lighting systems, efficient waste management, effective traffic control, and comprehensive environmental monitoring are few such applications which targets towards these objectives.

9.2.5.1 Utilization Scenarios

- Smart Lighting: The main aim of smart cities applications is to reduce wastage of resources by their optimum utilization. Energy is one such precious resource which must not be wasted as it has multiple direct and indirect effects. Smart lightning applications can help in saving energy by its smart utilization. It uses ambient light sensors, weather sensors, motion and activity sensors, connectivity through different standards and protocols to operate and adjust their brightness on the basis of real-world situations. It reduces energy consumption, saves money, improves public safety, and lower carbon footprint.
- Waste Management: In smart city, maintaining cleanliness is one of the top priorities. Use of IoT has changed the waste management process. It has made it efficient, cost-effective and environment-friendly. IoT sensors installed in waste bins can send updates to a central platform about fill-level, bin-location and temperatures inside. This information can be used to prioritize the filled bins and make

dynamic routing of collection trucks. It saves money and effort by reducing collection trips, fuel consumptions and labour costs. It keeps the city clean and reduce the emission of green-house gases.

- Traffic Management: IoT systems can be deployed to identify vehicle movement, road conditions, and traffic density in real time with the help of cameras, sensors and communication networks. This data can be used to measure traffic flow, detect congestions or accidents, optimize traffic signals and provide real time conditions to drivers. It results in reduced congestions, improved transportation efficiency, lower emissions, and safer driving conditions.
- Environmental Monitoring: Environment has direct effect on the residents of smart cities. Therefore, continuous monitoring of environment is of high importance. IoT sensors are deployed to continuously monitor the city environment. It helps city planners, environmental agencies, and public health authorities make informed decisions. PM2.5, PM10, CO_2, NO_2, SO_2, ozone etc. are measured with different sensors for identifying air quality. Noise levels, are measured with sound sensors. Water quality is also ensured by measuring pH, turbidity, dissolved oxygen, toxins, and contamination levels in different water resources. This data can be used for city planning, policy making, public-health decisions and sustainability initiatives. It increases real-time awareness of environment allowing proactive response in case of pollution spikes. It enhances public safety. It also allows long term environmental planning (Fig. 9.2).

Market Opportunities

With multiple nations taking initiatives, smart city market is expected to experience huge growth. UN focus for sustainable growth has fuelled the need for sustainable urban solutions. A survey by Allied Market Research, projects worldwide smart city market at a value of $6061.0 billion by the year 2030 with a CAGR 25.2% [8].

- Smart Infrastructure Solutions: Smart lighting, waste management, and traffic management, etc. are important contributors to infrastructure solutions for smart city. These all have substantial market prospects.
- Environmental Monitoring and Sustainability: Rising emphasis on sustainability, has brought the focus on companies that provide solutions for air quality, energy management, and water.
- Public Safety and Security: This one of the most basic demands. Any IoT application that will enhance public safety will have wide market reach anywhere in world. Intelligent surveillance, emergency response systems, and crime prevention technologies are notable IoT applications for public safety which attracts consumers of smart city sector.

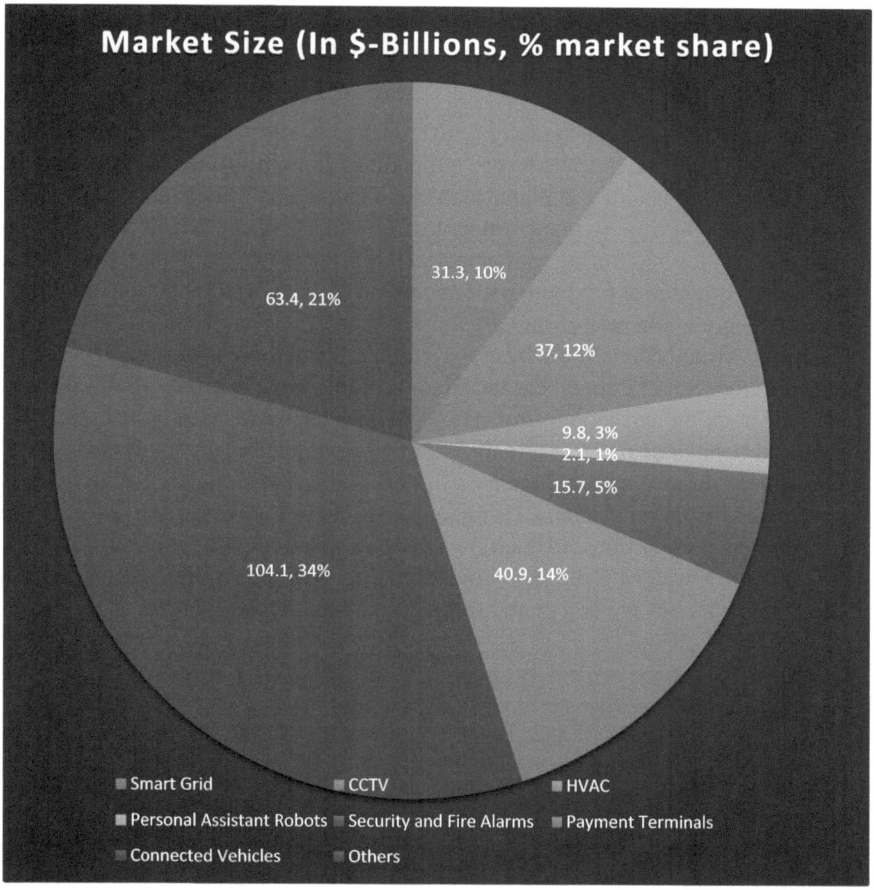

Fig. 9.2 Market Share of each sector [Statistica.com, https://www.itransition.com/iot/use-cases] [6]

9.2.6 Smart Agriculture

IoT has made major changes in the field of agriculture as well. Precision farming and livestock management are two prominent areas affecting the farmers where it has provided innovative solutions to monitor and enhance agricultural methods and to track livestock. This provides major opportunities for business as well.

Utilization Scenarios and Market Opportunities

The rise in population of world and increased urbanization has led to increase in food demand. IoT solutions can help in cutting down the operational costs and reduce wastage of energy. It will help in achieving the cost-cutting of whole agricultural process. Therefore, smart agriculture market is expected to achieve substantial growth. Based on a survey from Allied Research and Markets, the worldwide smart agriculture market is estimated to achieve a value of $29.2 billion by the year 2027 [9].

- Precision Farming: In this paradigm, IoT sensors are used to collect information like soil moisture and nutrient levels, temperature, humidity and weather patterns, crop growth leaf health and plant disease indicators etc. This data is analysed to find out the right time to plant, amount and timing to apply fertilizers, plan irrigation schedules and timing for harvest. All this leads to optimal resource usage, increased agricultural productivity, cost saving and environmental sustainability. It presents great potential to capture market share.
- Livestock Monitoring: Apart from agriculture, second most important resource of livelihood. Smart collars, ear tags and wearable sensors are used on livestock for their tracking and monitoring. These devices capture health indicators, locations and behaviour patterns. Analysis of these parameters are analysed to early detection of disease, to optimize feeding routines, achieve higher productivity.
- Automated Irrigation Systems: Using sensors to capture soil moisture and utilizing environmental conditions, irrigation valves can be automated. The schedules can be created for watering the fields by extracting patterns from real-time data. It minimizes water usage and guarantees optimal moisture and higher yields.

9.3 Summary

IoT is transforming industries and businesses by providing solutions which are more effective, and data-centric. Smart homes, IIoT, smart healthcare, smart cities, and agriculture are some of the diverse use-cases of IoT. Each use-cases offers distinct advantage regarding improvement in efficiency, reduction in costs, and improvement of user experiences.

With the ongoing advancement of IoT technology, there will be a rise in new applications and market prospects, which will in turn increase the adoption of IoT in many industries. Companies that possess the ability to create new and original ideas and provide dependable, adaptable, and easy-to-use Internet of Things (IoT) solutions will be leading the way in this technological revolution, gaining a substantial portion of the expanding IoT market.

References

1. S. Sinha, State of IoT 2025: number of connected IoT devices growing 14% to 21.1 billion globally, https://iot-analytics.com/number-connected-iot-devices/. Accessed 30 Nov 2025
2. SE1663, MarketandMarkets, RFID Market Size, Share and Industry Trends 2025 to 2033, https://www.marketsandmarkets.com/Market-Reports/rfid-market-446.html. Accessed 03 Nov 2025

3. SE 2909, MarketandMarkets, Industrial IOT Market Size, Share and Industry Trends, https://www.marketsandmarkets.com/Market-Reports/industrial-internet-of-things-market-129733727.html. Accessed 04 Nov 2025
4. ID: 978-1-68038-076-7, GVR Report on mHealth Market (2024–2030), https://www.grandviewresearch.com/industry-analysis/mhealth-market. Accessed 04 Nov 2025
5. Walid Miloud Dahmanea, Samir Ouchanib, Hafida Bouarfa, Smart cities services and solutions: a systematic review. Data Inf. Manage. **9**, 100087 (2025). https://doi.org/10.1016/j.dim.2024.100087
6. Smart Home-Worldwide, https://www.statista.com/outlook/cmo/smart-home/worldwide?srsltid=AfmBOordPUXyOl88Koz3iwzCoWYprO8GYr6_patp_l7KUI9m14s2a5co. Accessed 10 Nov 2025
7. Internet of Medical Things Market (2025–2030) Size, Share & Trends Analysis Report By Component (Hardware, Software), By Deployment (On-premise, Cloud), By Application, By End Use, By Region, And Segment Forecasts, https://www.grandviewresearch.com/industry-analysis/internet-of-medical-things-iomt-market-report
8. Report Code: A00290: Smart Cities Market Size, Share, Competitive Landscape and Trend Analysis Report, by Component (Hardware, Software, and Service) and Functional Area (Smart Infrastructure, Smart Governance and Smart Education, Smart Energy, Smart Mobility, Smart Healthcare, Smart Buildings, and Others): Global Opportunity Analysis and Industry Forecast, 2021–2030, https://www.alliedmarketresearch.com/smart-cities-market
9. Report Code: A05213, Smart Agriculture Market Size, Share, Competitive Landscape and Trend Analysis Report, by Type (Precision Farming, Livestock, Aquaculture, and Greenhouse), and Component (Solution, Service, and Connectivity Technology): Global Opportunity Analysis and Industry Forecast, 2021–2027, https://www.alliedmarketresearch.com/smart-agriculture-market

MIX
Papier aus verantwortungsvollen Quellen
Paper from responsible sources
FSC® C105338

If you have any concerns about our products,
you can contact us on
ProductSafety@springernature.com

In case Publisher is established outside the EU,
the EU authorized representative is:
Springer Nature Customer Service Center GmbH
Europaplatz 3, 69115 Heidelberg, Germany

Printed by Libri Plureos GmbH
in Hamburg, Germany